"When the Man waked up he said, 'What is Wild Dog doing here?'
And the Woman said, 'His name is not Wild Dog any more,
but the First Friend, because he will be our friend
for always and always and always.'"
—Rudyard Kipling, *The Jungle Book*

Australian Dingoes

DAWN OF THE DOG

The Genesis of a Natural Species

Janice Koler-Matznick

CYNOLOGY PRESS
Central Point
2016

DEDICATION

To *Canis familiaris* and all the individual dogs
that have enriched our lives.
—*Janice Koler-Matznick*

Editing by Alan Wittbecker, Bonnie C. Yates
Graphics and photographs: Janice Koler-Matznick, unless credited otherwise
Book Cover by Karen Adair, Janice Koler-Matznick
Book design and production by Karen Adair

Copyright © 2016 Cynology Press
All rights reserved under International and Pan-American Copyright Conventions. No part of this book may be reproduced in any form or by any means, including information storage and retrieval systems, without the prior written consent of the Author or Publisher.

Published by Cynology Press
Mail: 5265 Old Stage Road, Central Point, Oregon 97502 U.S.A.
Email: cynologypress@gmail.com

Publisher's Cataloging-in-Publication Data
Koler-Matznick, Janice
    Dawn of the Dog: The Genesis of a Natural Species / Janice Koler-Matznick
        p. cm.
Includes bibliographical references and index.

ISBN-13: 978-0-9974902-1-3 (cloth)
ISBN-13: 978-0-9974902-0-6 (paperback)
1. Dog. 2. Domestication. 3. Wolf.
4. Aboriginal Dogs. 5. Canid Ecology. 6. Ethology.
I. Title.

Early Reader Edition

CONTENTS

Acknowledgments 6
Preface 7
Glossary 10

Part I. Investigating Questions about the Origin of the Dog
Chapter 1. Once Upon a Time: Dog Origin Myths Debunked 14
Chapter 2. Dogma Revisited: Is the dog *Canis lupus*? 33
Chapter 3. Dog vs. Wolf: Natural Dogs and Gray Wolves............................ 49
Chapter 4. The DNA Story: Interpreting Genetic Dating 71
Chapter 5. Natural Dog Behavior? Free-ranging Dogs and Dingoes 91
Chapter 6. The Natural Species Hypothesis: Genesis of *Canis Familiaris* 133
Chapter 7. Conclusion 157

Part II. Examples of Primitive and Aboriginal Dogs
Introduction
The Dingoes.. 168
    New Guinea Dingo.. 173
    Australian Dingo 178
The Aboriginal Landraces 184
    Africanis.. 184
    Central African Forest Dog / Basenji... 187
    Canaan 191
    Formosan Mountain Dog 195
    Indian Native Dog 200
    Inuit Dog 205
Album of Aboriginal Dogs... 211

Figure Credits... 221
Bibliography... 223
Index 254
Author Information... 258

ACKNOWLEDGMENTS

The ideas and facts in this volume originated in the books and articles that shaped my ideas and understanding. Those authors supplied the pieces I put together in a new way. Authors that challenged my ideas kept me motivated to learn more in order to address their questions. Some authors graciously answered my questions and discussed ideas. The following were especially helpful, offering thoughts, data, or criticism (the most important factor in developing any hypothesis): Cheryl Asa, Adam Boyko, Vladimir Dinets, John W. Fondon, Michael W. Fox, Brian Hare, Ádám Miklósi, Darcy E. Morey, Ron M. Nowak, Stanley J. Olsen, Sunil K. Pal, Ben Sacks, Peter Savolainen, Wolfgang M. Schleidt, M. V. Sotnikova, Richard H. Tedford, Xiaoming Wang, Robert K. Wayne, Judith E. Winston, Stephen Wroe. I am sure I have missed mentioning many more.

I will be forever grateful to my family, friends, and acquaintances who listened to my ideas (ad nauseum), provided moral support when I was struggling, and stimulated my thinking with pertinent questions. My deepest gratitude is to my friend and colleague Bonnie C. Yates. She did what the best of friends do, and bravely offered constructive criticism, which helped me clarify my ideas. Bonnie's and Alan Wittbecker's excellent editing skills and helpful insights greatly improved the book. The book would never have been written without the help of my husband, Darwin Matznick, who took over the household tasks so I would have the time to research and write.

Of course I owe the most to the 23 dogs (including the 8 Rhodesian Ridgebacks I currently have) and 12 New Guinea dingoes who lived their lives as my friends, sharing my home. They were forgiving of my mistakes and tolerant of this human's slowness to recognize what they were trying to teach me.

Lastly, I would like to posthumously express my gratitude to writer Vickie Hearne who in 1999 during our discussions about the dog as a natural species said "You are sitting on one hell of a book."

Janice Koler-Matznick
Central Point, Oregon

PREFACE

It is a capital mistake to theorize before one has data. Insensibly one begins to twist facts to suit theories, instead of theories to suit facts.
— Sir Arthur Conan Doyle, "A Scandal in Bohemia" (1891) in *Adventures of Sherlock Holmes* (1892).

The dog is a wolf, but almost certainly not a domesticated gray wolf, *Canis lupus*, as commonly believed. This book presents the hypothesis that the dog is a distinctive biological species of wolf, an exceptional one with unique traits compared to other wolves. If this Natural Species Hypothesis is correct, the way we think about dogs will be fundamentally changed.

I never imagined myself as a heretic. I started out a true believer. For 30 years I was convinced the dog was a domesticated gray wolf. Almost everything I read about dogs said that the wolf was the ancestor of the dog. I assumed the experts who wrote about dogs had facts to back this up. Dissident ideas started creeping into my thoughts after I went back to college in midlife to study biology, and gained a greater understanding of how science is supposed to work. I learned the skill of critical thinking, to not accept things at face value, to question, and judge for myself how reliable a "factual" statement is. When I applied these principles to the idea that the dog is a domesticated gray wolf, it did not seem as credible any longer.

Here are the scientifically-proven facts about the origin of the dog to date:
1. The dog and gray wolf are very closely related.
2. The oldest recognized dog fossils are dated to about 14,000 years ago.

That is the current extent of hard, or unquestionable, evidence. The rest, including the concept I prefer, the Natural Species Hypothesis, is speculation based on inductive reasoning ("educated guesses"). The true, actual history of the origin and domestication of the dog may never be recovered. This leaves it up to each individual to examine the different ideas, carefully consider the evidence supporting those ideas, and then decide which they choose to accept as most likely. This book reports what I found over 20 years of investigation and explains why the Natural Species Hypothesis of the origin of the dog best fits the available evidence.

My not-a-wolf awakening came one day in 1992 while reading yet another account of the origin of the dog. When I got to the part that said 'because wolves are pack animals, if hand raised they easily fit onto human society,' I had my epiphany. I suddenly realized this had to be wrong. It occurred to me that nearly all of the authors of the wolf origin accounts either studied dead dogs and wolves (paleontologists, archaeozoologists) or modern domestic dogs (zoologists, ethologists). The closest most had ever been to a live wolf was probably observing them in zoos or from a distance in the wild. In my practical experience, I had raised a three-fourth wolf from 14 days of age many years before, and knew several people who had hand-raised wolves. None of these animals "fit easily" into human society. In fact, they were very difficult as companion animals because they were extremely destructive, hyper-reactive, and self-willed. The first presentation of my doubts about the wolf-origin idea, entitled "Why There Are No Wolf Acts at the Circus," was at the 1993 annual meeting of the Southwestern Anthropological Society.

My specialty is animal behavior, and it was my knowledge of the ecology and adaptations of predators, combined with my personal experience with captive wolves, that first made me doubt that Stone Age people could have kept them successfully as companions or neighboring mutualists. These doubts led me to explore the literature in subjects relevant to the prehistoric period when dogs must have been domesticated: paleoanthropology, gatherer-hunter lifestyles, and canid paleontology. Eventually, I expanded my inquiry to the process of domestication in various species, to the questions about different ways to define species and their relationships, and to other subjects that related to when, where, and how the dog may have originated. I became convinced of three basic fundamentals: (1) The wolf is not the direct ancestor of the dog; (2) The human and dog relationship started well before 15,000 years ago; and (3) This relationship began when the dog voluntarily attached itself to human society.

Today, after more than two decades of investigation, I feel confident saying that the majority opinion is not based on thoughtful consideration of the evidence supporting a wolf origin, because I could find no such analysis. Instead, it appears that the dogma of wolf origin is faith-based. Most scientists have failed to question its underlying assumptions because they are comfortable with the dogma. Faith and dogma should have no place in science, but scientists are human, subject to human weaknesses. When some plausible idea has become entrenched in science, the evidence required to change a faith-based belief has to be nearly irrefutable to be seriously considered as an alternative. While the Natural Species Hypothesis is not presently accepted by most, there is nothing currently known that directly contradicts it.

The idea the dog could be a natural species has been mentioned many times in the last two centuries, most notably by T. Studer (1901), F. E. Zeuner (1963), H. Epstein (1971), M. W. Fox (in a 1973 article for the American Kennel Club Gazette and later in his 1978 book), and C. M. A. Baker and C. Manwell (1983). Based on the body of secondary evidence I have collected, I think they are correct. It is entirely possible that the dog originated from a natural species of wild dog, a close relative of the wolf. This book provides a critique of the wolf-origin hypothesis and an explanation of the evidence supporting the Natural Species Hypothesis alternative. Like everything about the origin of the dog, beyond the two facts listed above, the natural dog hypothesis still needs to be supported by more hard evidence, and the Wolf Origin Hypothesis, currently based on the genetic closeness of the dog and wolf, must be more critically evaluated.

In Part I, Chapter 1 examines the two most popular dog origin stories: The Pet Hypothesis and the Self-Domestication Hypothesis. It also discusses why Stone Age people would be highly unlikely to allow wolves to co-habit with them. The most important reasons are that wolves that associate people with food are dangerous, and that they require a high protein diet, which would be a huge burden for Stone Age people to provide. Chapter 2 discusses the proposition that the wolf and the dog are the same species. It explains where the skulls of wolves and dogs differ and offers biological reasons for the differences. The dog's distinctive skull traits may be better explained as adaptations inherited from the ancestral dog's unique lifestyle than by self-domestication, domestication or artificial selection.

Chapter 3 explores the question of the dog's scientific name and physical differences between the dog and wolf, some rarely mentioned or previously dismissed as "merely" domestication effects. Chapter 4 discusses some recent dog and wolf genetic studies, specifically questioning the reliability of genetic dating, and how the assumption that the gray wolf is the direct ancestor of the dog might affect the conclusions. In Chapter 5, I examine the natural behavior of dogs as exemplified by the aboriginal village dogs of Africa and India, feral dogs in Italy, and the Australian dingo. My concept of the origin of the dog and speculations about the ancestral dog species, and where I think it originated, are covered in Chapter 6. The Conclusion summarizes Part I and talks about on-going research that has yet to be published.

Part II describes and pictures the two most primitive dog races still in existence, the wild Australian and New Guinea dingoes, and some of the ancient races of domestic dog, the aboriginal village dogs. It was while looking at village dogs and dingoes that I became convinced the dog is not a wolf, because they are all variations on a basic type, which is nothing like a wolf.

For those who want to delve deeper into a topic, each Chapter is supported by references and additional information in the Notes section. Every source directly referenced is in the Bibliography along with a selected sample of related references I have examined in my research. All photos used in the text are credited to the copyright owners in the Figure Credit section. The Index contains a list of subjects and important names that appear in the text.

This science-based book will not be an "easy read" for some. It is meant to inform, not merely to entertain. Scientific concepts critical to the discussion are briefly explained and when technical terms are necessary a translation in plain words is provided. My hope in writing this book is that it will help open up minds to consideration of *all* evidence about the origin of the dog. Until new discoveries in genetics, paleoanthropology, and paleontology become available to support either the Wolf Origin or the Natural Species Hypotheses, the only reasonable approach is to reserve final judgment until additional evidence is available. Meanwhile, the hypothesis that either has the most indirect support, or violates the fewest established facts, should be given preference. Those who read this book thoughtfully will have the essential information to decide for themselves which hypothesis is the most logical and well supported.

GLOSSARY

The terms used for dogs can be ambiguous, because words can have variable functional definitions. Knowing the author's definition is vital to understanding the material presented. In this book, the following important definitions are intended.

Aboriginal. Of or relating to the people and things that have been in a region from the earliest time.

Aboriginal village dogs. Ancient indigenous populations of dogs, also called "landraces" for their adaptation to the local environments, whose movements and reproduction have been under no or very little direct human control.

Adolescent dog. Dogs from 6 months to about 18 months of age.

Artificial. A process or form that exists primarily due to human effort.

Breed. An artificial subpopulation of "purebred" domestic dog created and maintained by human-directed selection.

Canid. Any of the taxonomic family Canidae, carnivorous digitigrade animals that includes the wolves, jackals, foxes, coyote, dingoes, and the domestic dog.

Dingo. Wild subspecies of dog, indigenous to Australia (*Canis familiaris dingo*) and New Guinea (*Canis familiaris hallstromi*).

Dog. The species *Canis familiaris*. Unless otherwise noted, when used alone "dog" refers only to the natural generalized dog, such as the dingo and aboriginal village dog.

Domestic dog. All members of the species *Canis familiaris* except the dingoes.

Feral. Having no social tolerance of people, either never being socialized to humans or having lost their trust of people.

Free-breeding. Dogs that choose their own mates.

Free-ranging. Not confined; roaming free without any direct human control.

Mixed breed. Crosses between modern domestic dog breeds; dogs of no discernible purebred ancestry. SYNONYM: **mongrel**.

Natural. Existing in or formed by natural processes; not cultivated or altered by direct human action.

Niche. An environment that has all the things that a particular animal needs in order to survive. For canids the niche includes their method of hunting and the size of their typical prey.

Primitive. Seeming to come from an early time in the ancient past; closely approximating an early ancestral type; not derived. Synonym: original.

Puppy. Dogs from birth to 6 months of age.

Stray. A dog of modern derived populations that is socialized to people but roaming unsupervised.

Wild. Not domesticated; living (or capable of living if captive) in the natural environment as an integral part of the ecology.

Wolf. Canids belonging to the genus *Canis* that have relatively robust skulls and teeth compared to other species in the genus; when used without a modifier here indicates only the holarctic species *Canis lupus*, the gray wolf. If referring to a specific population or subspecies of *Canis lupus* or to a separate species of wolf, an adjective is added, as in "paleolithic wolf" or "Indian wolf."

An adult female New Guinea dingo.

PART I

Investigating Questions about the Origin of the Dog

CHAPTER 1. ONCE UPON A TIME ...

There is no great harm in the theorist who makes up a new theory to fit a new event. But the theorist who starts with a false theory and then sees everything as making it come true is the most dangerous enemy of human reason.
—Gilbert Keith Chesterton, *The Flying Inn* (1914, p.103)

There are countless stories concerning the origin of the dog. They all assume the animal started as a gray wolf and was changed to "dog" after domestication. The proposed details of how and when wolf became dog vary considerably, but there are only two basic underlying premises for the most well accepted ones. Either the wolf was domesticated by man, or the wolf self-domesticated by becoming a voluntary scavenger-around-humans. This Chapter is titled "Once upon a time ..." because the vast majority of these dog origin stories have about as much basis in fact as other common cultural myths. There may be some truth in them, but that golden kernel is often devalued by the profuse chaff of the highly unlikely human and wolf behaviors described.

These stories are "plausible explanations" that do not directly contradict anything known, so they have been accepted without much questioning and the main elements handed down to the present with only minor variations. Discussion of the fallacies in the stories and specific examples follow the sketches of the two main versions.

VERSION I. "Once-upon-a-time" Dog Origin Story: Purposeful domestication (or The Pet Hypothesis).

Long ago, when humans still made a living by gathering and hunting, they adopted young wolf cubs as pets. Because the wolf's nature is to live in a group and submit to a leader, some of these cubs became tamed adult wolves adapted to being part of a human pack. The humans desired wolves because they were useful. They helped locate game and trail wounded animals for the people. They also held large animals at bay, helped bring them down, and acted as protective guards of the human camps. The "friendly" camp wolves that acted more puppy-like as adults were favored by the humans. This is how the wolf became the dog, and why dogs, like wolf pups, have smaller teeth, higher foreheads, and less slanted eyes than adult wolves. Later, when dogs were used as guards for livestock, they were also selected for the up-curved tail and odd non-wolf colors, in order to easily tell them apart from the wolves that might attack the flocks.

A specific example of Version I comes from J. Scott and J. L. Fuller (1965), who were some of the first to scientifically study dog behavior. Note that in 1965 it was thought the dog originated only 9,000 BC (or 11,000 years Before Present; equivalent to "years ago from 1950"):

> As to how domestication took place, we can only guess. Probably it happened very simply. . . . Primitive peoples everywhere in the world frequently adopt young birds and mammals as pets. We can suppose that wolves hung around the primitive agricultural villages of Europe scavenging any waste food or bones that were thrown away, and that the human inhabitants might frequently come across wolf cubs in the spring. . . . One can imagine a wolf puppy growing up in a village, fed at first and later existing on scraps. As wolves and dogs still do today, it became adopted into human society and established a territory around its home. Its sensitivity to the approach of strange animals and people at night must have been immediately valuable. Later, when goats were domesticated, its dog-like descendants could warn their owners of wild wolves, which might attack the herd. *(p. 55)*

The earliest reference I found for this Pet Hypothesis was by F. Galton in his 1883 treatise on domestication. The purposeful domestication idea has recently fallen out of favor with biologists, who now adhere to the more biology-based Natural Selection Hypothesis in which the wolf turns itself into the dog.

VERSION II. "Once-upon-a-time ..." Dog Origin Story: Self-Domestication (or the Garbage Dump Hypothesis).

Long ago, when humans first settled down in permanent villages, a pack of wolves was attracted to a village by the refuse dump. Because they kept the garbage cleaned up and served as watch animals, the humans did not object to their presence. The less fearful wolves, the ones who could best tolerate the close proximity of humans, were able to scrounge the most food and so produced more offspring than the fearful wolves. Thus, generation by generation the proto-dog village wolf population became naturally (genetically) tame. Because scavenging does not provide as much nutrition as hunting, natural selection favored smaller body size in the proto-dog, and over generations resulted in reduced size compared to wild wolves. In addition, their relative tooth size (size of the teeth in relation to body size) went down quickly, because proto-dogs no longer needed big teeth for hunting.

A specific example of Version II comes from R. Coppinger and L. Coppinger (2001), who developed their hypothesis based on the idea that it was highly unlikely Mesolithic people produced the dog from the wolf by artificial selection. They say:

> The outline goes:
> People create a new niche, the village.
> Some wolves invade the new niche and gain access to a new food source.
> Those wolves that can use the new niche are genetically predisposed to show less "flight distance" than those that don't.
> Those "tamer" wolves gain selective advantage in the new niche over the wilder ones.

> In this model, dogs evolved by natural selection. The only thing people had to do with it was to establish the villages with their attendant resources of food, safety, and more opportunities for reproduction, which provided the naturally tamer wolves with increased chances of survival . . . . For simplicity, let's call this new niche the town dump. *(p. 57)*

The Coppingers go on to conclude:

> Those village-oriented canids also began to change shape. Their new shape made them even more efficient at scavenging. The scavenger wolf was beginning to behave and look dog-like. Besides evolving tameness, it acquired a size and shape that were specialized for scavenging—a smallish size, with a proportionately small head, smaller teeth, and just enough brain to point it in the right direction. These wolves were fast becoming adapted to the niche, and were incipient dogs. . . . The smaller animal, the dog, is a new form adapted specifically for the new niche, the dump. *(p. 61)*

The fallacies of this explanation for the physical differences between dogs and wolves are addressed in the next two Chapters.

FALLACIES

The current understanding about human cultural evolution is that permanent year-round settlements first arose about 12,000 BP in the Middle East, and by then there were domestic dogs.[1] It was only after this that plants and other animals were domesticated. Wild plants such as millet, wheat and barley may have been encouraged or purposefully sown for harvest, and some wild tubers may have been cultivated even before permanent settlements. Prior to 12,000 BP most people lived a semi-nomadic gatherer-hunter lifestyle within defined territories. They moved camp according to seasonal availability of prey and food plants. Their shelters, which only needed to be used for at most several weeks, were frequently made of skins, bark, or thatch. This makes locating and studying the remains of these cultures difficult. Bones left over from meals were often discarded on the surface, where they rapidly disintegrated due to weathering or were demolished by animals and fungi. It is generally believed that such a lifestyle restricted bands to only about 25 to 50 members, although there may have been seasonal gatherings of several bands into larger groups. All these factors, plus the low level of effort to find sites in many areas of the world, contribute to the scarcity of secure knowledge about humans of the period.

Given what is known about the state of human culture at the time man and dog joined up, and current understanding of wolf behavior, there are several unlikely assumptions in the above dog origin stories. Foremost among them is the idea that wolves and humans could co-habit. Humans at the time dogs and people must have partnered up had only hand-held stone weapons to defend themselves against large carnivores. Long distance weapons like spear throwers and bows appeared about 30,000 BP, which is likely, in my opinion, after humans and dogs began their association, so people were potentially easy prey for wolves. Wolves back then had no reason to fear humans any more than they feared other animals with the ability to defend themselves, such as elk or horses. Even armed adult humans would not have been able to defend themselves from a wolf pack any better than, say, a deer could with its horns and hoofs. A single person with a club or spear could not successfully defend herself against more than one wolf. A lone person with arrows would have to be exceptionally fast and accurate to fight more than one wolf. Children would be especially vulnerable to predation. Would these people allow, let alone encourage, wolves to live around their camps? It is amazing that the Fiennes, in their book *The Natural History of Dogs* (1968), actually say "No doubt on occasion wolves would attack stragglers or seize young children, and man was probably to some extent in awe of his rival predators...." yet they still propose that wolves would have been accepted as camp followers.

FIGURE 1.1. *The old tales of Red Riding Hood (also called "Red Cap" in some European countries) and the Big Bad Wolf were based on the every-day reality of the times.*

The authors of both versions of the wolf origin stories, and people who accept them, have a much romanticized concept of wolves. They obviously lack in-person, hands-on experience with wolves and do not understand carnivore ecology. They misunderstand wolf behavior and make assumptions that are unrealistic. The Fiennes are correct that wolves prey upon humans when given the opportunity, and are also correct that wolves and humans were direct competitors for large game. Both facts, however, make a human-wolf partnership extremely unlikely. Wolves that are regularly hunted with guns do learn to avoid humans, and pass this learned fear on to their offspring. Once they are no longer hunted, though, wolves will prey on humans. In India children are regularly seized and often killed by wolves. They even have a name for this behavior: Child-lifting. In the recent past in rural Russia and other European countries wolves, especially in winter when other prey were scarce, killed people, mostly women and children. Recently, in Canada and Alaska, wolves accustomed to humans as a source of food (either through direct feeding or through scavenging at refuse dumps) have attacked and killed people.

In his book *Wolves in Russia* (2007), W. Graves mentions that in the reports he read of Russian wolves attacking humans, those few cases of "pet" wolves that escaped were especially dangerous to people.² This makes sense, since they are not afraid of humans, have no experience hunting normal prey, and associate humans with food. V. Geist, a well-respected Canadian wildlife biologist who edited Graves' book, contributed instances where wolves had lost their fear of humans and subsequently attacked and sometimes killed people.

Geist became interested in the risk wolves pose to humans while living on rural Vancouver Island in Canada where, between 1990 and 2003, a local pack of wolves started acting less fearful of humans. The human population had expanded significantly into rural areas in recent years and the never-hunted wolves were accepted as part of the local ecology. Locals believed the wolves were not a threat, and they were not, at first. But, over time the wolves became bolder, staying closer to the houses, hanging out on the edges of yards, and trailing people in the woods. As someone who knows wild animal behavior, Geist recognized the signs of a predator learning about and testing potential prey. The wolves were carefully studying their potential prey, humans. The pack on Vancouver Island eventually escalated the testing into actually nipping at people.

Wolves are cautious and assess potential prey for defenses by observing them for an extended period. The next step is testing the prey, which includes approaching it to within touching distance to see if and how it fights back. If it runs, it is prey for sure, and the wolves will attack.³ So, when wolves began acting "friendly" and "playful" toward people, Geist warned the locals the wolves had become much too familiar with humans. Eventually 13 wolves were killed, and no humans were harmed in that area. However, in 2000 two wolves from another Vancouver pack that had become food-conditioned in a camp ground attacked a camper while he slept, causing serious injuries before he was saved by fellow campers. In 2005, a university student at a research station in Saskatchewan, Canada, was killed while out on a solo hike by wolves that had been coming to forage at the station's dump where they had been photographed. M. McNay (2002), an Alaskan wildlife biologist, researched wolf attacks in North America and found that between 1970–2000 there were 13 attacks (not all resulted in human injuries).

The common denominators of situations in which wolves become dangerous to humans are, as provided by Graves and Geist:

1. Severe depletion of natural prey.
2. Wolves find alternative food sources among human habitations (purposefully fed, scavenging garbage, or preying on pets and livestock).
3. Wolves have not been hunted with guns.

Understandably, to gain public acceptance for wolf conservation and restoration, wolves needed to be portrayed as not dangerous unless rabid. The implication is that only wolves out of their normal minds would attack humans. The cases of attacks mentioned here, however, were not by rabid wolves. In Eurasia, where many cases of rabid wolves biting people were recorded, the attacks by rabid wolves are clearly different than predation attacks. Rabid wolves are disoriented, hyper-agitated, slashing and biting at everything that moves, often injuring several people in a village before dying or being killed. The attacks above were by perfectly normal wolves practicing perfectly normal wolf behavior.[4]

WOLF BEHAVIOR DIFFERENCES

Like all animals, including humans, the personality and temperament of wolves varies across a natural spectrum from extremely shy to extremely bold.[5] Because the inborn tendencies interact with many external variables during personality development, not every wolf, even bold ones, will become dangerous to humans. Some argue that we should not extrapolate prehistoric wolf behavior from the behavior of wolves recorded over the last few hundred years, because wolves back then could have had very different behaviors. Usually what they intend to convey is the idea that wolves back then may have been less dangerous, and so could be more trusted around people. That seems highly unlikely. The opposite is much more realistic: Prehistoric wolves would be less fearful because they were not targets of purposeful human hunting using powerful, automatic, long-distance weapons. They would more likely consider humans direct competitors. As we have seen, lack of fear of humans is the first step toward predation on people. Therefore, prehistoric wolves would probably have been even more inclined to attack humans, not less. Surely, then as now, the boldest most experienced wolves would lead the others during hunts and test the potential prey's defenses, presenting the most danger to people. The shyest wolves, those that tended to be the most cautious and prone to fearful reactions, would have been the safest to allow around humans, but least likely to try out the new niche (a term from ecology that basically means the habitat an animal chooses to live and feed in and its relationships with other organisms in that habitat).

While it is not outside the realm of possibility that, as in the Pet Hypothesis, wolf pups could have been caught before five weeks of age when fear to unfamiliar things emerges and raised in the camps for a period of time, there are many factors that make it extremely unlikely prehistoric people would live with mature wolves. Many of these same factors, discussed below, also apply to the Natural Selection idea.

The natural behavior of any species only changes when the environment changes or some individuals migrate to another environment where new selective pressures favor a different behavior. An environmental factor could be climate change, massive fire, and for predators depletion of preferred prey or the arrival of a new kind of prey. Animals migrate into new types of habitats because there is no room for them in preferred, known, habitat, or by chance (chance displacements usually only happen in smaller species like lizards, mice, rats, bats, and birds).

It is well accepted that in the wild predators that are direct competitors for the same prey, such as wolves and humans, never cooperate. They either fight for access to prey or keep away from each other. To avoid each other, they may partition the resources up, possibly by living in non-overlapping ranges or hunting at different times. They also may directly interact, as when the larger predator or a pack of predators, such as wolves, take kills away from smaller or solitary predators, like coyotes (or coyotes from foxes and so forth).[6] The incentive for large predators like wolves to join forces with another large predator like humans would have to be great enough to overcome all kinds of natural behavioral barriers.

If the Natural Selection story (Version 2) is accepted, then we must ask, what was desirable enough for the wolf to voluntarily change most of its natural habits? Perhaps if the local prey base was decimated by some natural disaster, such as a huge fire, hungry wolves still young enough to adapt could conceivably see the scraps left by humans as "worth" putting themselves in danger from a competitor and possible predator. However, gatherer-hunter groups, except those in extremely prey-rich areas or who hunt very large animals, leave little uneaten around their camps. If the DNA dating of the wolf/dog separation at 12,000–35,000 BP is accepted (the uncertainties are discussed in Chapter 4), there were no "garbage dumps" of left-over food like those in later permanent settlements. Killing large game with stone or bone-tipped hand held weapons is hard and dangerous work, so it only makes good sense to utilize everything you can of each animal, including the fat-rich marrow of the larger bones. According to estimates done by experts, wolves need an average of about three pounds of meat a day just for maintenance level activity in captivity, and more for growth and reproduction.[7] Even two tamed, dependent wolves would therefore need to eat the equivalent of half a deer a week. Domestication is an evolutionary process (meaning genetic changes over time) that takes at least several individuals reproducing over dozens of generations. Wolves take a long time, a minimum of two years, to become efficient predators if they have the example of older experienced wolves to learn from. It would be very time-consuming, perhaps impossible, even for an experienced trainer with modern methods, to teach a young wolf to hunt large game properly.

One of the few attempts I know of to train human-raised wolves to hunt on their own was related by L. Crisler in her 1958 book *Arctic Wild*. She describes raising and filming a litter of five hand-raised, free-ranging wolf pups in remote Alaska for a documentary. The Crislers intended to let the wolves go free at the end of the project, so they tried to get them to hunt the abundant caribou in the Alaskan wilderness. They failed, and ended up having to take them home to Colorado to live in pens (their sad fates were chronicled in her 1968 sequel book *Captive Wild*). Unless the prehistoric people were willing to hunt for their wolves, the wolves probably would still have to mainly depend on what they could catch for themselves. Without pack cooperation this would have been small game like rabbits, birds, or reptiles. If they were ranging out far enough from the camps to kill their own food (most of the small game near the camps would likely quickly be taken by the people), they probably came in contact with resident wild wolves, which could either breed with them or attack them. The latter is more likely unless a lone estrus female wolf or proto-dog met a lone male wolf or proto-dog, or vice versa.

PARTNERSHIP LIMITS

Prehistoric people would have had to have a good reason to keep a pack of expensive tamed wolves, tolerating their often destructive and dangerous (especially to children) behavior. What use would wolves, or even proto-dogs many generations from wild wolves, have been 20,000 or more years ago? Various ideas about how wolves "benefited" a hunter-gatherer band have been offered. None, however, have much support from the literature on the behavior of captive wolves.

Some think that the human-wolf partnership started when humans took advantage of the wolf's superior hunting skills by capturing and raising them to assist with hunting large game. Actually, wolves are no better at hunting than people. Wolf hunts end in failure much more often than success. A similar idea is the humans followed the tamed wolves when they hunted, and then took their kills from them, thus using the wolves to get their own meat. This idea is based on modern dog behavior. In India, if the locals notice free-ranging pariah dogs catching a deer or other good-sized prey, the people easily chase the dogs away and appropriate the kill. Trained to hunt birds and small game for humans, modern dogs readily give up anything they catch, even bringing it to their handler. Dogs are also used to bay large prey like moose, holding it in place until the hunter arrives to dispatch it. However, both pariah dogs and hunting dogs are many thousands of years away from their ancestors, and during that time dogs have been selected for being cooperative with humans. Taking away small prey from a free-ranging primitive dog, such as a dingo, or from a wolf, would be impossible. They would merely keep carrying it away from the person. Try taking food away from a tame captive-raised wolf and you are likely to find it is not so easy.

FIGURE 1.2. *A wolf defending its kill, possibly, given the upward angle of its gaze, from an approaching human or upright bear.*

Chapter 1: ONCE UPON A TIME

Wolves, even very young ones, have a strong biological imperative to defend their food. A hand-raised wolf will bite the hand that feeds it to protect its food. I know that from personal experience and from other captive wolf keepers. The bite would probably be inhibited, just hard enough to convince the person (as it would another wolf) to not try again to take the food. But even an inhibited wolf bite could easily leave a human with bruises and puncture wounds. Perhaps proto-dog camp wolves could be beaten off of big prey by a large group of people using clubs and sticks, held at bay until the carcass was butchered, and then allowed to have the unwanted parts as a reward. Without the reward, the tame proto-dog wolves would soon stop chasing prey when humans were around, as they could not benefit from it. This concept of humans exploiting the incipient dogs is the only version of the human-wolf "cooperation" scenario that is at all convincing. The relevant question is, would the humans find it was more work to follow the wolves, often at a fast trot for long distances, and confiscate the prey and defend it, than just doing the hunting themselves? If the answer is yes, then keeping the tamed wolves to hunt for the people would be more hassle than they were worth.

The least realistic version of the hunting benefit scenario states that wolves and humans first joined forces because they could hunt together better than either could alone: Mutual benefit. For instance, wolves could trail wounded game for the humans, or locate game the humans could then kill. The supposed benefit to the people would be an increased hunting success rate. The benefit for the wolves would be the energy savings and reduced chance of injury because humans make the kills. This idea of course goes directly against what is known of the behavior of predators competing for the same prey, as outlined above. It also shows ignorance of the exceptional hunting skills of hunter-gatherers still evident in primal people today. Surely those in the past would have been at least as competent. These people need no help to locate or track prey, wounded or otherwise. Aboriginal hunters have nearly unbelievable skills for reading animal signs and tracks, and they know from experience where and when the different prey species are likely to be located. They can tell from a few hoof or paw prints when the animal passed, how fast it was going, and even sometimes what sex it was.[8]

A weakness of the idea of human and wolf cooperation is that wolves, unlike dogs, are very resistant to inhibition training. They are difficult to train to respond reliably to commands such as "Stay" or "Wait."[9] The Australian dingo, which was traditionally captured from the wild as a young pup and raised in the camp, can serve as a surrogate model for the tamed wolf proto-dogs. The Aborigines used tamed dingoes to hunt small game in forests, because the dingoes can locate hidden prey in holes and rocks, and so are useful there. However, when the intended quarry was the large kangaroos out on the plain, the Aborigines threatened the dingoes, threw rocks at them, to make them stay in camp. If the dingoes followed the men and sighted kangaroos, the dingoes rushed off after them, chasing the game away from the hunters.[10]

FIGURE 1.3.
A wolf shadows a bear with vigilant interest. Neither poses an immediate danger to the other or to any offspring, and they are not in competition over food, so there is temporary reciprocal tolerance (unless the bear decided not to move on).

Another fallacious benefit often proposed is that tamed wolves were useful as watch or guard animals, protecting the camp and people from other predators or hostile humans. Scott and Fuller (1965) said: "As wolves and dogs still do today, it became adopted into human society and established a territory around its home."[11] The implication is that the wolves then protected their territory (area around a seasonal camp). Even if wolves lived within a camp, they would make poor "watch wolves" because wolves do not bark a warning alarm, merely snort through their nose, and become alert and agitated when they detect something unfamiliar coming their way. Then they retreat to a safe distance to observe the potential threat and evaluate it. If the people noticed the nervous behavior of the wolves, they would be warned something strange was coming, but unless that "maybe dangerous" something turned out to be a smaller predator or an unfamiliar wolf, the camp wolves would not attack it to "defend their territory." If the something was a large predator such as a bear, the wolves might harass it from a short distance and convince it to leave while avoiding its claws and teeth. It is more dangerous to them to have a large predator living nearby than to make an attempt to get it to move on. But, if the larger predator stands its ground, they would merely wait for it to leave.

The first biological imperative for all wild animals is to avoid serious injury, in order to live to reproduce. This is why most wild animal mothers make only cautious efforts to save their young attacked by predators. This can be seen in multitudes of nature documentaries. In North America there are reports from biologists that when they crawl into wild wolf dens to examine pups the parents are distressed and nervous but do not try to attack the biologists. The dingoes were not of much use as guards to the Aborigines because they were never fully changed from natural wild behavior (dingoes split off from the main population of dogs several thousand years ago) and they never developed a warning bark. They are suspicious of unfamiliar humans but do not "defend" the aboriginal camps. Companion New Guinea dingoes also have never demonstrated any guarding instincts except concerning mates and puppies. If they are not well socialized, they may growl and show stress and anxiety when strangers enter their territory, but they will avoid contact without offensive aggression.

I have not found any reputable account of wolves ever being fully integrated into human society. In some places, such as Mongolia, wolf cubs are raised in captivity so they can be tethered to prevent their escape while being hunted by tamed eagles or dogs. This is a traditional way to train eagles and dogs to hunt wolves. Even though raised by people, these tamed wolves are not integrated as adults into the social fabric of the people, because wolves are very difficult to live with after puppyhood and must be carefully managed and strictly controlled.[12]

Only young wolves are kept by primal peoples because wolves, especially adolescents, are impulsive and can be incredibly destructive. Even playing with a person, a wolf can damage them with teeth and paws. Unlike wolf play partners, humans do not have thick fur and skin to protect them from playful bites. Another reason adults are not kept is that when approaching their first breeding season after sexual maturity wolves may begin "testing" pack members, including those same-sex humans they consider part of their pack, in order to define their status in the group. This "testing" consists mostly of posturing and threatening, but if not handled appropriately it can escalate into attack. Normal wolves inhibit their attacks on familiar wolves, and presumably would for familiar humans, but humans are fragile creatures compared to another wolf. Even inhibited bites can result in serious injuries. Today, with antibiotics and medical care, injuries from wolf social competition probably would not be life-threatening, but during the Paleolithic they would have been.

All of these drawbacks, plus the fact that most of the "benefit" reasons given for humans adopting wolves are unrealistic, create doubt about wolves and humans co-habiting. To "see" this, visualize the following scenario about a tamed prehistoric wolf, and then decide for yourself if it is likely our ancestors integrated wolves into their society, or even allowed them to hang about the camp. Keep in mind that 20,000–40,000 years ago wolves, even older puppies, could not be contained and so were free-ranging around the camp. Metal fabrication had not yet been invented: There were no chains. Tethers would have been made of leather or plant fibers, which any wolf with even a full set of milk teeth would make short work of.

> **SCENARIO:** *A dozen dome-shaped grass huts stand in a broad meadow next to a fast-flowing creek. The warm late spring afternoon air is filled with the buzz of insects. At a fire pit in front of one of the huts a woman bends over, taking large roots from the basket at her feet and pushing them into the coals around the fire. Glancing up, she sees her daughter, just recently walking on her own, stumbling towards her, smiling, still carrying the deer rib bone the mother gave her an hour ago to sooth the child's sore gums. The girl extends her arms toward her mother. Just then a young wolf, already larger than the child, races past the girl, grabs the rib bone, knocking the child down, and gambols off with its prize.*

How long do you think parents would tolerate this kind of behavior before the wolf was killed or banished?

So, the following reasons for the wolf being domesticated do not seem realistic.
1. Sentry and guard service.
2. Hunting partner for large game.
3. Garbage clean up (little edible waste before permanent settlements).

ANCIENT DOG SKULL STORY

The oldest skulls, ones all scientists agree are domestic dogs, are about 12,000–14,000 years old *[see Note 1]*. These early dogs are much smaller than any known wolves, including wolves that lived in the same time period. The only exception is the extinct Japanese wolf *Canis lupus hodophilax,* which was supposedly only about 18 inches tall at the shoulders (island animals are often smaller or larger than continental cousins). Older "dog" skulls described recently are questionable and may have been merely a different type of wolf. These include a skull from Belgium, dated to about 31,000 BC, and Siberian skulls from about 31,000 BC and 21,000 BC. These specimens were about as large as the local gray wolves of the period found in the same sites. They were designated "dog" only because they have rather short, broad muzzles compared to known wolves (the diagnostic skull traits separating the dog and wolf are discussed in Chapter 2). Sticking to the wolf origin of the dog hypothesis, the authors considered the broad muzzle with slightly crowded teeth a domestication effect. However, the skulls could have belonged to a wolf adapted to a different niche than the contemporaneous gray wolf. The 10,000 years separating the oldest and youngest of these very similar Paleolithic skulls makes the different species idea the most likely, as wolves under domestication would surely have acquired more "dog" traits in that long time period. Recently, mtDNA (mitochondrial DNA, inherited separately from the nuclear DNA and only from mothers) analysis of these skulls showed they have haplotypes (DNA variations) not found so far in either dogs or wolves, although closely related to both. If the "domestication effect" proposed for these skulls is correct, the process of taming and domestication of their ancestors had been on-going for a long time by 31,000 BC. Interestingly, both the wolf and "dog" skulls found at these sites had openings in the cranium indicating they were eaten by humans.[13] These Paleolithic short-faced wolves are discussed in more detail in Chapter 3. The most frequently given dates for the separation of dog and wolf, based on DNA sequencing, are 15,000–20,000 BC (DNA dating is dussed in Chapter 4). Given these dates, all authors agree on one thing: The change from wolf to dog must have started when humans still lived primarily as gatherer-hunters, well before the first known permanent settlements, which date to about 12,000 BC. The next oldest domestic animal, the goat, does not show up in the archaeological record until about 9,000 BC, so the dog was definitely fully domesticated long before any other animal.

DOG ORIGIN TIMING

Another thing to think about concerning the usual time estimate for dog origin is that at about 12,000 BC (the oldest uncontested archaeological dog) dogs are not found in just one geographic area, but many places in Europe, Asia and Japan, and by at least 10,000 BC in North America. How long would it have taken for the dog to spread all over the Northern Hemispheres via trade and human migration? While no direct estimate is possible, the ancestors of the Australian and New Guinean aboriginal people arrived 45,000 BC, only about 5,000 years after modern humans supposedly first left Africa. So this world-wide dog expansion could have taken about the same time as the first population expansions of modern humans, i.e., 4 or 5 thousand years.

Because of the chance factors of preservation and search effort involved in paleo-archaeology and paleontology research, the earliest dated find of something is a minimum date for its origin. The actual maximum date, or "true" date, for the origin of a species or cultural innovation may never be known. There are unquestionable dog skulls from about 12,000 BC, so it must have evolved into identifiable shape prior to that. If the degree of differences between wolf and dog were present in two groups of other similar mammals, the skulls would likely be determined to come from separate species. It is reasonable to assume that under the Natural Selection model of dog origin, the new scavenger-around-humans niche the dog's ancestral wolves (if the gray wolf is the dog ancestor) adopted would take the same amount of time to cause change in bones and teeth as any other population adapting to a new niche. Unfortunately, we may never have certain knowledge about rates of evolution at the species level, how long the transition from A (wolf) type to B (dog) type takes under natural selection, but surely hundreds to thousands of generations. Perhaps the dog is much older than archaeology and genetics are indicating.

So, why are there no dogs in older archaeological sites? The simple answer is probably we have just not found them yet, and may never locate any. The only sources of information we have about human life 20,000 plus years ago are based on artifacts from archaeological sites and observations of the few gatherer-hunter people who kept their traditional lifestyles into the recent past. The Upper Paleolithic archaeological record, the period from about 12,000–50,000 BC, is very sparse and no confirmed dogs have been reported in that period that are older than about 13,000 BC. The early Upper Paleolithic is the time in which dogs and humans must have begun their association, but the record is very incomplete because the people of that time made camps that were used perhaps for only a few days to a few months. They left little behind except stone flakes and in some places middens (places at camp sites where domestic refuse accumulated) and bone piles. At that time, man had no metals, pottery, or glass, and so those things that survived into the present, buried under protective dirt or ash or in caves, are mostly bone or stone. If they did not eat dogs on a regular basis and discard the bones, especially skulls, which are more likely to be preserved, in pits or large middens, the rare dog remains would not survive. Until recently, tools and some bones were all that were available to tell what these people ate, and what the local environment might have been like. The paleoenvironment, the local vegetation, amount of rain fall, and average temperatures, can be worked out from the animal and plant species present on a site by extrapolating their preferred habitat and food from similar modern species.

FIGURE 1.4. *(below)* A generalized Human Cultural Timeline relevant to the human-dog relationship: The Upper Paleolithic, Mesolithic and Neolithic (the Paleolithic era—Stone Age—began with the appearance of anatomically-modern humans in Africa 2 million years ago). Dates are approximate, and not all stages of culture occurred in all areas. When a range of domestication dates is given, these are from F. B. Marshall et al. (2014). Single dates are the most common used in the other sources consulted.

Upper Paleolithic: 50,000–15,000 BC (all dates BC, or years ago)

45,000– Modern humans colonize Europe; Western, Southern and S. E. Asia; Australia/New Guinea

40,000– Neanderthals become extinct (or nearly so); earliest known cave paintings

36,000–31,700– Paleolithic "wolf dogs" in areas now known as Siberia, Czech Republic and Belgium

35,000– First known depictions of humans in sculpture

32,000– Oldest estimated (by genetic dating) separation of the dog and gray wolf

28,000–20,000– Harpoons, needles, saws, pottery (figurines) invented; fiber is used to make clothes and nets; the last maximum stage of the Ice Ages

25,000– Huts made of rocks and mammoth bones located in today's Czech Republic are first known permanent structures (may have been seasonal)

Mesolithic: 25,000–10,000 BC

20,000– Oldest pottery storage/cooking vessels (China)

17,000–13,000– Paleolithic "wolf-dog" in central Russia.

15,000–10,000– The mega fauna of the Ice Ages (cave bears, cave lions, saber-cats, Irish elk, and many other very large species) go extinct

13,000–10,000– End of last glacial period

13,000–11,000– The Natufian culture in the eastern Mediterranean has the first known permanent settlements pre-dating agriculture; age of oldest dog skulls

12,000– Most common genetically estimated date for separation of dog and gray wolf

11,000– Neolithic Revolution (agriculture/farming) begins in the Near East and quickly spreads as far as Europe

11,000– Dogs are buried with humans in the Eastern Mediterranean Natufian culture

11,000–9,000– Goat domesticated in Western Asia

12,000–10,500– Sheep domesticated in Middle East/Western Asia

12,000–8,300– Pigs domesticated in Middle East/SE Asia and China

10,500–7,500– Cattle (two species) domesticated in Middle East and Southern Asia

Neolithic: 10,000 to about 4,500 - 2,000 BC

10,000–9,000– Barley and wheat cultivated in Mesopotamia (northern Iraq)

10,000–8,000– End of the last Ice Age: Sea level rises to present position, land connecting SE Asia and Indonesia (Sahul) and other low-lying land

8,500– Cat domesticated in Fertile Crescent of Middle East/Egypt

7,500– First evidence of copper smelting in what is today Serbia

5,500– Horses domesticated in Central Asia; the chicken in SE Asia

5,200– Egyptian dynasties begin

In the last few decades, new archaeological techniques such as pollen analysis and improved microscopic and genetic analysis of bone remains, plus careful ethnographic studies of remnant gatherer-hunters, have broadened our knowledge of prehistoric environments and possible lifestyles. Prehistoric cultures were more complex than previously thought. For instance, we now know that early humans were more creative than formerly imagined. Recently a 500,000 year old freshwater mussel shell with zigzag engravings on it, done before the shell was fossilized, indicate *Homo erectus* could think abstractly, and a Neanderthal cross-hatch rock engraving in a Gibraltar cave dated to at least 39,000 BC proves those early people also had a symbolic element in their culture.[14] We know that one of these archaic humans or early modern humans began our symbiosis with the dog, which has greatly benefited both humans and dogs. However innovative and artistic these people were, imagining a future relationship with a commensal predator (wild, but adapted to living around and among humans) and shaping it to fit their needs would have been beyond their capability. Until they settled into permanent villages, no animals could adopt a commensal niche, except the dog, which somehow found a way. Hopefully new discoveries will someday reveal more about the early dogs.

OLD MYTHS

In addition to the main origin story versions, there are several minor myths surrounding the origin of the dog and of certain breeds that need to be examined. I will only address a few here. One recurring myth is that people, especially Northern people such as the Inuits, have regularly and purposefully bred their dogs to wolves in order to "strengthen" some quality of their dogs. Genetic testing has revealed that some other domesticated species, such as donkeys and camels, were purposefully bred back to their wild ancestors and even to other close wild relatives, to keep the domesticated populations strong for their jobs as draft animals and physically well adapted to the harsh environments where these domesticants originated.[15] This is highly unlikely for dogs. While the early dogs undoubtedly sometimes bred with wild dogs and wolves, after full domestication hybridization with wild canids was rare (this is discussed further in Chapter 6). In their book *The Natural History of Dogs* (1968) the Fiennes discuss, without any caveats, Aristotle's claims that dogs were tethered out to be crossed with wolves, and offer the example of the crossing of Indian pariah dogs ("notorious for their savagery") to dholes (a distant dog relative now in danger of extinction) or to the small Indian wolf. The Indian pariah dog, an ancient landrace described in Part II, is not only *not* "notorious for their savagery" but, on the contrary, are known to be shy toward humans, and although they will kill wild game, they can easily be trained to leave livestock, even chickens, alone. There are no verified reports of dholes crossbreeding with dogs, and none of the local pariah dogs breeding with the Indian wolf.
The reality is that while living in the same areas for thousands of years, and having the same breeding season, Indian wolves and native aboriginal dogs do not hybridize (this has been proven with genetic testing), a strong indication that the dog and wolf are separate species. Genetic tests of Indian wolves so far have not turned up any dog genes, although an mtDNA haplotype in Basenjis and other dog breeds has been interpreted as deriving from the Middle Eastern wolf, so there may have been a rare hybridization in the past.[16] R. and L. Coppinger, who have carefully studied sled and livestock dogs, debunk the myth that the Northern native peoples of the Americas crossed their sled dogs with wolves to "improve" them. They describe the behavior and morphology a useful sled dog must have and point out that crossing to a wolf would produce useless hybrids, or worse than useless, as they would be a detriment to the team. It took thousands of years of special selection to make the various types of dogs that are useful to humans. Crossing with wolves only sets the line back to the beginning, and human lives depend on the performance of the dogs. When dogs are a survival necessity no sane person would ever purposefully engage in this folly.[17]
The Fiennes book (1968) also perpetuates the old myth that Native Americans tamed

local wolves and coyotes and used them in the creation of their own type of dog. In discussing the migrations of Paleolithic Asian people into the Americas, they say:

> It is known ... that wherever these Mongoloid tribes ... went they domesticated whatever wild wolf or wild dog was present in the vicinity. In the northern regions, timber wolves were domesticated and from these various local breeds were developed.... Those tribes that penetrated farther south domesticated the prairie wolf [coyote] ... from which one breed of America dog is derived. *(p. 24)*

However, although some still believe this just-so story, in fact there is no evidence Native North Americans domesticated or kept captive wild canids of any species, except the foxes of the California Channel Islands.[18] This separate domestication idea is pure speculation. All of the dogs so far excavated from Native American pre-European contact sites are fully recognizable as domestic dogs (see Marion Schwartz's *Dogs of the Early Americas, 1998*), with no discernible relationship to any wild canid. One possible exception is a single canid skull which may have been a wolf hybrid.[19] Recent DNA studies have proven that the Xoloitzcuintli, a native Mexican dog known to have been in Mexico for at least 2,000 years, and the Inuit dogs, have only DNA related to the Eurasian dogs tested.[20]

The Fiennes, like some other authors, also confuse reports of tamed wild canids with the origin of domesticated dogs. They claim:

> In South America there exist a number of forms of forest-living wild dogs [meaning dogs as in the general term canids] from which sprang domesticated dogs. These are ancestral to the various Azara dogs, which are found in many countries of the South America continent and even in the Falkland Islands. Another form whose puppies are captured and reared in captivity is the strange-looking small bush dog *Icticyon (Speothos) venaticus* of Brazil. Races of man at an Upper Paleolithic cultural level or its equivalent thus possessed an urge to provide themselves with some form or other of domesticated dog. This is strong evidence for the proposal made in this work that our modern breeds have derived in the main from at least four different ancestral stocks. *(p. 24)*

While the indigenous people probably kept individual animals as "pets" for entertainment and as educational opportunities for children, they surely already had dogs descended from those their ancestors brought into the New World. Newly acquired canids would be tamed, but not domesticated, and not useful except as novelties. In any case, the South American wild canids are not close relatives of dogs and wolves. They are more distantly related to domestic dogs than coyotes and jackals and no dogs are part South American canid. Chapter 4 addresses the recent studies of dog DNA not available to the Fiennes that prove beyond much doubt that domestic dogs are descended from one ancestor species, with rare hybridization with Old World and perhaps New World gray wolves.

SUMMARY

Not all dog origin stories are equally supported by documented evidence. Many authors have not carefully researched their subject, or their work was written before some significant contradictory evidence was available. It is always appropriate to listen to hypotheses about the dog's origin with a skeptical and questioning mind. Stories should be compared, and what they offer as evidence for their version verified by independent sources. No dog origin hypothesis can at this time be proven beyond reasonable doubt, but some are more logical and are supported by several lines of evidence and reasoning. That said, I will begin the support of my hypothesis that the dog is a true natural species in Chapter 2 by providing evidence that indicates that the dog, *Canis familiaris*, is not the same species as its assumed ancestor, the gray wolf, *Canis lupus*.

NOTES

1. Tchernov E, and F. F. Valla, 1997; Zeder, M., 2012.
2. Graves, W. N., 2007; Mech, L. D. and L. Boitani (eds.), 2003a.
4. McNay, M., 2002. In North America wolves, having been severely persecuted and hunted by early trappers, farmers, and ranchers, were until recently rare and their natural prey plentiful. Now wolves are protected in many areas, and humans are populating what remains of natural wolf territory, including increased use of wilderness parks and reserves. It is therefore imperative that all of us have a realistic understanding of normal wolf behavior, and ensure that wolves are not purposefully fed or allowed to scavenge at dumps, because this will create dangerous conditions for humans. In areas were wolves range, the public should be educated to report bold wolves, and garbage dumps in wolf territory should be fenced to prevent habituation. In order for wolves and humans to coexist, those wolves that learn to prey regularly on livestock and pets should be taught through non-lethal means this is a dangerous behavior. If this fails, they unfortunately need to be culled. Wolves should be preserved in large tracts of wilderness, and some predation on livestock far from human habitation tolerated for the benefit the wolves have on the local ecological balance. But if they are allowed to work up to attacking people or killing pets and livestock near homes, general sentiment will surely turn once again to extermination.
5. Svartberg, K., 2007; Yli-Renko, M., O. Vesakoski and J. E. Pettay, 2015.
6. Berger, K. M. and E. M. Gese., 2007; Paquet, P. C., 1992.
7. Mech, L. D. and L. Boitani (eds.), 2003a.
8. Lee, R. B. and R. H. Daly (eds.), 1999; Churchill, S. E., 1993.
9. Moore, J., 2005.
10. Meggitt, M. J., 1965. However, D. F. Cahir and I. Clark (2013) related stories

from early European residents, mostly repeating stories told by the native locals and not from their own experience, that said in some areas in Australia dingoes were used to hunt large kangaroos. Having experience with dozens of New Guinea dingoes raised from birth by experienced people and carefully trained and socialized, I seriously doubt these reports that dingoes were useful as hunting aides for large animals that can run (or hop) rapidly for long distances—more rapidly than humans can run. Even with modern expert behavior modification and training, unconfined tame dingoes (like captive-raised wolves) cannot be voice controlled as they are instinct-driven and highly independent; they decide when to cooperate, and most of the time, going for the bird in the bush prevails over capitulating to human commands. The early proto-dogs associating with humans who had never trained animals would surely have been no better at cooperating. It takes several dingoes to chase and bring down a large kangaroo, and by the time the Aborigines arrived at the kill site, the kangaroo would undoubtedly have been well chewed up. If the people did not care about the meat being damaged, this might be acceptable. Then dingoes could be used to hunt kangaroos and the carcasses co-opted by the people, one scenario proposed for humans and wolves.

11. Scott, J. P. and J. L. Fuller, 1965.
12. Wilde, N., 2000.
13. Germonpré, M., et al., 2009; Germonpré, M., M. Laznickova-Galetova and M. V. Sablin, 2012; Morey, D. F., 2014; Sablin, M. V. and G. A. Khlopachev, 2002; Thalmann, O., et al., 2013.
14. Joordens, J. C. A., et al., 2014; Rodriguez-Vidal, J., et al., 2014.
15. Larson, G., et al., 2014.
16. Gray M. M., et al., 2010.
17. Coppinger, R. and L. Coppinger, 2001.
18. Witt, K. E., et al., 2015.
19. Walker, D. N. and G. C. Frison, 1982.
20. Leonard, J. A. et al., 2002.

CHAPTER 2.
DOGMA REVISITED: IS THE DOG CANIS LUPUS?

"The dog, beloved as humankind's most faithful companion, descends from the wild gray wolf, but how this happened has been a matter of conjecture, controversy and confusion." —Acland & Ostrander (2003, p. 201)

Nature is messy, always changing, evolving, never static or completely predictable. The living things are the messiest part of Nature, especially when you look closely. And the absolute messiest things in biology are "species." Species are like clouds. Looked at one way they appear to be this, and viewed from a different perspective they are that, or becoming that. Clouds have "species" (categories of related types) too, with names like cumulus, cirrus and stratus. But, like biological species, the edges of these types are fuzzy and they sometimes interact, combine, and separate again. Many volumes have been written about the "species problem" in biology; that is, how to recognize and define species, and the debate shows no sign of resolution satisfactory to all biologists.

Knowing a bit about how species are defined and named is important for understanding why the idea that the dog is a wolf has become so entrenched. Currently, there are at least six species concepts in regular use that have different (but often related) criteria defining what is included or excluded from the species category: The Biological Species Concept, the Phylogenetic Species Concept, the Ecological Species Concept, the Recognition Species Concept, the Evolutionary Species Concept, and the Hennigian Species Concept. There are even more concepts in minor use, such as the Chronospecies Concept used by paleontologists. Why are there so many? Because biologists working on different problems in ecology, conservation, genetics, and other fields, need definitions that are relevant to the work they are doing. They see species in different ways. Plus, all Nature is in continual flux. The ancient saying from Heraclitus that you can never put your foot into the same river twice is true. Species are clouds, and where you place the boundaries, what you include within the species depends on when, where, and how you look at them. However, a truly "good" species is considered a species under most of the concepts using various defining criteria, and the majority of biologists agree on them.[1]

A species can be thought of as a sort of collective "individual" because the organisms that belong to it behave in many respects as a unit. For instance, it is the species that evolves, not the separate organisms. Species also have perceptible boundaries that interact with other species, like our own permeable skin that interacts with everything inside and outside us, exchanging molecules. Like individual humans, these packets called species have unique "life" histories, and the individual organisms that make them up, like our own cells, have histories separate from the whole. Both the whole and the parts live, reproduce, and die. When a species reproduces, the parent species may continue and perhaps produce another new species again someday, or become extinct.

Species reproduce by splitting in various ways. Sometimes some of the individual organisms start (for whatever reasons) to eat a different food, or to be active at a different time of day. Because the population exhibiting the new behavior is then more likely to meet and therefore breed with each other than with the rest of the species, over time the two populations evolve along different paths until they accumulate enough differences to be recognized as separate species. This is "sympatric speciation."

Sometimes a small group of organisms becomes isolated from the main population in some way, usually by a geographic barrier that prevents interbreeding (a river changes course and separates the two populations; a climate change makes the area between the two populations unsuitable for them to cross; or, a few individuals get transported on a raft to an island. Over time the isolated population changes enough that it is morphologically (physically) or genetically considered a separate species. This is "allopatric speciation" and is believed to be the most common way new species are formed.[2]

Species must have "diagnostic" traits, some characteristics that all of the individuals that belong to that species have that is specific to them (well, at least nearly all in mammals, because in mammal species there will always be some individuals who lack some trait, like a red fox without the usual white tail tip). In mammals, diagnostic traits can be just about anything physical, physiological, or behavioral.[3] For example, the identifying differences between two similar species can be body size, breeding season, coat pattern, courtship behaviors, or different sets of DNA mutations. Taxonomy, the discipline devoted to identifying species and their relationships, can be a specialty, a sideline, or a hobby, but in all cases it is, if done properly, tedious and exacting work. To be reasonably certain the specimens one has examined belong to a unique species, the taxonomist's task is to look at all the relevant similar species, by examining them in person or through the literature about them.

The first person to describe species gets to choose its scientific name. Scientific names were invented by Carl Linnaeus in 1735. Back then, the belief was that species were God's creations of "special kinds," and species were thought to be fixed, unchanging. However, Linnaeus devised a way of officially naming species that provided information about which other species it was most similar to in appearance (nothing was known then about genes or how species are related). The "scientific name" has two parts and must be unique. No two species of the same class can have the same name. The first part, the name of the genus the animal belongs to, is capitalized. The genus may contain other species recognized as very closely related, and species can be added or subtracted from the genus as new information is discovered that says they do or do not meet the criteria for that genus. Today the critical "yes" or "no" information usually comes from a combination of genetic analysis and/or a large sample of individuals that reveals previously unrecognized variability of some species in the genus that was wrongly given a different genus name. This system of naming still works pretty well today.[4]

The second part of the name, not capitalized, is the species epithet. Both the parts of the name should be italicized. Today, newly named species must have a published written description or illustration of at least one individual of that species that is detailed enough others can identify additional members of that species. This was not traditional before the mid-twentieth century. This first named individual is called the "type specimen." So, scientific names have specific meanings and at least in modern practice refer to an original type specimen. They are meant to convey the most accurate information available about where that species fits into the bigger scheme of similar species, like a Dewey Decimal System library catalog number for a book. The species can be moved from one genus to another, or changed from a species to a subspecies or vice versa, and the species epithet usually stays the same in order to connect that name back to the original description.

Technically, the first time a scientific name is used in a publication, it is by tradition supposed to be followed by the name of the person who named it (or re-named it) with the date the name was published. The attribution is, however, regularly omitted because most people interested in the author's work would know what species they mean, and would never bother to look up the original description. The minority of the time for attribution to be vital is when someone wants to know how the original author described the species, what criteria were used, in order to decide personally if it is a good species, if it is in the right genus, and if the name was properly assigned.

Linnaeus named 4,236 species of animals, and two of them were the dog and the gray wolf *(See Figure 2.1)*. In his book *Systema Naturae* (1758) he made the dog, not the wolf, the "type" species (a typical species representative of the genus) for the genus *Canis*. He named the dog first, *Canis familiaris* Linnaeus, 1758, then a page later, the gray wolf, *Canis lupus* Linnaeus, 1758. Today the *Canis* genus includes the wolf (all species), dog, jackal (three species), and the coyote. By all criteria, including genetics, this genus is a closely related group: They are more closely related to each other than any of them are to other canids (other members of the "dog family," Canidae, which includes foxes, dholes, African wild dogs, and several less known living and extinct species).

Now back to the opening question: "Is the dog *Canis lupus*?" In other words, is the dog a gray wolf, the species named by Linnaeus based specifically on the Holartic (Swedish) gray wolf, as opposed to just some type of "wolf-like canid"? Actually, the question could be turned around. The dog is the type species for the genus *Canis*, and the dog was named a page before the wolf in the same volume (if two names are being used for the same species, the first one published has priority), so if the two are the same species, then we could be asking: Is the wolf a dog, *Canis familiaris*? The latter may seem inappropriate to those who are convinced that the dog is a domesticated version of the wolf, and so believe the wolf is the "parent" species, but if the two are considered the same species, which species name they should be called is a legitimate, although peculiar, biologically-based question. The rest of this Chapter will explain why I think the answer to both questions is "No."

DEFINING SPECIES

The most stringent and popular species definition today is from the Biological Species Concept (hereafter BSC) first published by Ernst Mayr in 1942 and later slightly revised.[5] In general, the BSC states: "Biological species are groups of interbreeding natural populations that are reproductively isolated from other such groups." The emphasis in this definition is not on the degree of morphological (physical) sameness or difference, but rather on inter-relatedness through breeding. This is the only definition most non-biologists are ever exposed to, and then in a very simplified version that states that two true species cannot produce fertile offspring, and if they do interbreed and have viable offspring, then they are not true species. The usual example provided is the mule, the infertile offspring of a cross of a horse and a donkey. Recent information from DNA analysis of many species has revealed hybridization is much more frequent than previously thought, and usually does not result in sterile hybrids.[6] Being a "good" (acceptable) species under the BSC definition just means that species living side-by-side in the same area can each maintain their general over-all genetic uniqueness, despite a few individuals making mistakes in choosing a mate. In other words, when the two

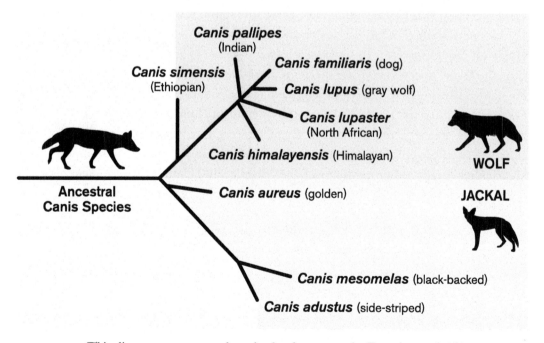

FIGURE 2.1. *This diagram represents where the dog fits among the Eurasian and African species in the genus Canis. The length of the lines between the species indicates how closely they are related to each other, with all C. lupus represented by one designation.*

species meet in nature, some may hybridize, but the two populations do not meld into one universal gene pool.

Why, then, is it now such a common belief, both in the scientific community and the general public, that the dog is *C. lupus*? This has always been a prevalent idea among biologists because the dog and wolf are so similar in general body plan, and the public has picked up on this belief. For instance, the dog and wolf have the sturdiest teeth and skulls in the genus, with very similar shapes, and these are major characters (traits) that often indicate two species either are closely related or that they have very similar lifestyles (in this case the robust teeth and skull that indicate they both can hunt large prey). Then, when the first preliminary genetics studies were done in the 1990s, they showed that the dog and wolf are extremely close genetic relatives. The just-so story seemed supported and became more entrenched.

Still, most of those biologists who think the dog probably is a descendant from the wolf rather than merely a close relative consider the dog different enough to deserve its own species name, the name it was known by for over 200 years. Changing scientific names is a serious matter, as the whole system is set up in order to provide continuity through time. If a name is changed, the author is supposed to publish the reason for the change, such as new evidence that indicates that it is not a proper species, or it belongs in another genus. There is an actual organization of experts on names, the International Commission on Zoological Nomenclature (hereafter ICZN), which maintains a rule book on valid ways to name species and acts as an international panel to decide controversies over names. However, since these are merely community rules, compliance is voluntary.

The idea the dog is a wolf became more well-established due to a name change made by W. C. Wozencraft, the author of the canid section in a very popular book listing mammal names, Mammal Species of the World, edited by D. E. Wilson and D. M. Reeder and published in 1993 (and subsequent editions) jointly by the American Society of Mammalogists and the Smithsonian Institution. These two organizations are greatly respected for their high standards of scientific practice, and those using the book often assume that it represents the most accurate determination of species' names, based on general scientific consensus. Unfortunately, this is not necessarily correct. Trying for a consensus on thousands of names would be a monumental task, especially because sometimes biologists have different opinions about them. So, each section is written by an expert on those kinds of mammals, and what names they provide is based on their expert opinion.

Many species have been named more than once, either by mistake (the second author did not know it had been named before, or thought that it was a different species than the one already named, but later it was found to be the same as the first named), or when evidence shows it should belong to another genus, and so forth. If there is more than one name for a species they are called synonyms. A list of all of a species' names is a "synonymy" and is supposed to include data about the names in the synonymy, including the author, date of publication and reference to the publication it was named in. This list should also include any recognized subspecies. A scientific name, per the rules, can have a third term that designates it as a subspecies (more about these later). In his *Canis lupus* synonymy list in *Mammal Species of the World*, Wozencraft renamed *Canis familiaris* as a subspecies of wolf, *Canis lupus familiaris*. Why? Because by the ICZN rules, the domesticated form of a species whose ancestor still exists should not be given a name separate from that ancestor.[7] Wozencraft's justification (from the 2005 edition) for this change was not very satisfactory. He said only: "*Includes the domestic dog as a subspecies, with the dingo provisionally separate—artificial variants created by domestication and selective breeding (Vilá et al., 1999; Wayne and Ostrander, 1999; Savolainen et al., 2002). Although this may stretch the subspecies concept, it retains the correct allocation of synonyms.*"[8] So, no objective "justification" was given for calling the dog a subspecies of wolf other than Wozencraft believed the dog is an artificial variant of wolf created by domestication and listed as references three papers on dog genetics, which opined that the dog is *C. lupus*. Given the early state of knowledge of genetics related to speciation at the time, this was very poor support for changing a species name that had been in use for over 200 years.

The genetic papers referenced by Wozencraft do show that *C. lupus* and *C. familiaris* are very close genetic relatives, and each states that, therefore, the dog is a domesticated *C. lupus*. However, close, even extremely close, genetic relatedness, does not necessarily indicate there is an ancestor-descendant relationship. In any case, calling the dog a subspecies of wolf does not merely "stretch" the concept of a subspecies as defined by Ernst Mayr in 1970, it violates its basic premises of genetic isolation due to geographic isolation and genetic mixing due to interbreeding when the subspecies is again connected to the main population. Declaring the dog a subspecies in name only was a way to maintain the *familiaris* to differentiate dogs from the supposed ancestral type, link it back to the original description by Linnaeus, acknowledge the general belief in the wolf origin for the dog, and satisfy the ICZN's rule that a domestic species should have the same name as the living ancestor. It was not good science.

The modern definition of a subspecies can be summarized as: A subspecies occupies a separated geographic area from other populations of the species, and differs in some significant trait from other populations of the species. Subspecies are basically detached spatial population units, divided from the rest of the species by a geographic barrier of some sort, e.g., mountain range, large river, or unsuitable habitat, that have evolved distinguishing traits, which can be used to identify the subspecies reliably. These diagnostic characters are the same types used to tell two species apart: DNA, and physical, physiological, or behavioral traits. Subspecies are in fact "incipient species" that if left isolated to continue their separate evolution, eventually may become a true species (in the allopatric speciation mentioned before). The understanding has always been that if the geographic barrier was eliminated, a subspecies would begin interbreeding with the main population and the two would merge, because in isolation inherent reproductive barriers between the parent population and the subspecies had not yet evolved.

So, even if the wolf was proven beyond doubt to be the dog ancestor, the domestic dog does not meet the criteria of a subspecies. The wolf and dog gene pools have maintained their general integrity over thousands of years of coexistence. For example, the comparisons of wolf and dog mitochondrial DNA [mtDNA], which is inherited only (well, almost always) through the mothers, and Y-chromosome types inherited only through the male line, shows that out of hundreds of samples only a handful of wolf and dog types are the same.[9] There is no doubt that, in maternal evolutionary lines at least, dogs and wolves are more closely related to each other than either is to any other species, and the genetic study results are discussed in Chapter 4. The few shared mtDNA types may be the result of occasional dog-wolf hybridization, although they could be inherited from a recent common ancestor. Due to the close relationship, it is no surprise there is occasional natural cross breeding: There is infrequent hybridization even between the more distantly related wolves and coyotes (*Canis latrans* Say, 1823), and hybridization between Ethiopian wolves (*Canis simensis* Rüppel, 1835) and domestic dogs is posing a threat to this endangered canid. The surprising thing is how rare dog-wolf hybridization must have been, because if it was common there would be many more shared wolf and dog haplotypes. The similarity in mtDNA is not evidence they are the same species, and in fact the rarity of shared mtDNA and Y-chromosome types indicates they are two good biological species that have maintained their genetic integrity in the face of thousands of years of potential contact. Thus, even using the BSC, the most stringent species definition, *C. lupus* and *C. familiaris* are not the same species. Currently, only about one third of scientists are using *C. lupus familiaris*, with the rest preferring the dog's original name.

Earlier, similarity of the dog and wolf skull was given as one of the reasons it is now such a common belief that the dog is a wolf. These two taxa (the plural of taxon, which means a distinctive population or group at any level of scientific classification, usually referring to subspecies or species) have the sturdiest teeth and skulls in the genus *Canis*, with similar shapes, and such similarity often indicates the two taxa are closely related or have very similar lifestyles. The differences that can tell a dog from a wolf have been dismissed as by-products of the domestication process, often attributed to "just" differences in developmental timing (discussed further in Chapter 4). The rest of the Chapter looks at the differences and the possible reasons for them.

COMPARATIVE STUDIES

To date, no one except me and my colleague Bonnie Yates has included dogs in comparative studies of canid skulls and teeth—studies that could determine the dog's possible adaptation to a particular lifestyle. Our results, discussed later, indicate the dog does not match any known niche (lifestyle) for any extant canid species. It has been considered inappropriate to include dogs in comparative studies of natural wild canids, because most think the dog is merely an artificially changed descendant of the wolf and therefore dog morphology (physical shape) is due to human influence. Domestic dogs are probably the most variable domestic animal, with dwarfs, midgets, giants, and an array of skull shapes that is astounding, varying from virtually no muzzle to muzzles relatively as long as a wolf's ("relatively" means in comparison to another body measurement, in this case overall skull length). This extreme variation has put researchers off, but these outlying extremes should not be included in a comparative study. To find out if the original dog's skull was adapted to a certain lifestyle it would be vital to use the most natural dogs, those with the least artificial selection for appearance. The best comparison would use the dingoes, which have escaped 4,000-12,000 years of artificial selection (depending on the dating method used) compared to domestic dogs and are the only dogs that still exist as full-time predators. This is what we did in our study. The next best dogs for comparison would be the aboriginal land race dogs, like those in Part II; dogs from populations that have continuously existed as free-breeding pariah or village dogs for the last 12,000 years or more. The third best subjects would be the indigenous sled dogs, the Inuit dog, Greenland dog, and Siberian husky, because they have been fairly pure evolutionary lines for a couple thousand years and were selected for work, not for appearance.

Two major studies comparing dog and wolf skulls, by R. Wayne and D. Morey, greatly influenced how dog skulls were viewed. Both concluded that dog skulls are most similar to puppy wolf skulls, that dogs are wolves stuck in a juvenile physical developmental state (neotenous). They came to this conclusion because dogs have relatively short, wide muzzles, broader craniums, and higher foreheads than adult wolves; they are similar to puppy wolves. Their conclusion was bolstered by R. Coppinger's behavioral research, which concluded that dog behavior is equivalent to juvenile wolf behavior. In fact, after these and similar conclusion by scientists, it became popular to say the retardation of skull development was a result of dogs being selected for tame and submissive behaviors, neatly tying the two things together as cause-effect. These conclusions were accepted without serious criticism because on the surface they seemed reasonable. However, there are several problems with these conclusions, which should have been questioned, and probably would have been if this research had been done on any other species besides the wolf and dog. For instance, it was never explained just how selection for tame behavior translated physiologically into major changes in the skull or why the rest of the dog skeleton did not become juvenile.[10] There were also technical shortcomings in the actual research that are too involved to address here, such as the type of samples (dogs) included and the fact that they did not include the jaw. The main problem was none of the researchers even considered the alternative, that even before domestication started the dog may have been adapted to a lifestyle different than the wolf, and that is why their skulls are different. Why didn't they look at alternatives? It did not occur to them because they never questioned the wolf as the direct, recent ancestor of the dog. They set out to answer how the wolf turned into the dog rather than objectively studying the dog, and that was the answer according to the current state of knowledge.

In her 2011 paper about the results of her study of heterochrony of the dog skull (heterochrony is changes in the rate or timing of development, like those responsible for paedomorphism [retention of juvenile traits in the adult], that lead to changes in adult size and shape), A. G. Drake reported that dog skulls are not like puppy wolf skulls. She was the first to examine a large sample of dog and wolf skulls, newborn to adult, and found that adult dogs do not have the same skull shape as wolf puppies. She states flatly "Dogs are not paedomorphic wolves." And "[T]he heterochronic model fails to explain the evolution of domestic dogs from wolves."[11] For instance, the dog's palate tilts at an angle to the face, while in the puppy and adult wolf, face and palate are on the same plane. Tilting of the palate would not be an effect of just developmental timing.

The following discussion about the specific differences of dog and wolf skulls and jaws is illustrated with the skull of an adult female Minnesota wolf and an adult male German Shepherd-husky mix dog, both of which are in the collection of the National Wildlife Forensic Laboratory in Ashland, Oregon. Minnesota wolves are in general somewhat smaller than most Northern gray wolves, and females are usually more refined than males and so closer to dogs of the same size in dimensions and tooth size. The dog skull is "normal" without any exaggerations. These skulls were chosen because they are only a few millimeters different in over-all length, and neither had defects such as injuries, making them directly comparable.

Remember that when the term "dog" is used here it refers only to the generalized type with average characteristics similar to those of the most ancient races of dog like the dingoes, and "wolf" only to the Holarctic gray wolf. The main diagnostic differences are that dogs have, relative to the wolf:

- Shorter, wider muzzles
- A higher rise above the eyes (higher foreheads)
- Wider, more rounded craniums (brain cases)
- Less robust teeth with more rounded canine teeth
- Smaller and flatter bullae (the bony cases surrounding the inner ear bones)

Dog eyes also sit at a less acute angle than wolf eyes. The orbital angle indicates the degrees between a line through the horizontal center of the eye and a level horizontal line touching the inner corner of the eye. It is greater than 50° in dogs and generally less than 50° in wolves, often close to 45° *(see Figure 2.2. below)*. In addition, the horizontal ramus (lower jaw) of the dog is convexly curved beneath the carnassial teeth (the largest meat cutting teeth) while the wolf's is usually straight. The top of the vertical blade of the back of the dog's jaw is curved backward a bit, while the wolf's normally is not. Of course, as in all things biological, there is variability in the characters, and in some cases certain dogs and wolves will not meet all of the criteria. Sometimes a wolf will have flatter bullae or a curved jaw, or a dog will have a relatively narrow cranium. Therefore, the determination if the skull is of a dog or a wolf depends on several traits taken together.

The bones above the eye sockets are called frontals. They are higher in the dog because the dog has much larger frontal sinuses, or cavities, inside those bones. All species in the genus *Canis* have some small sinuses there, but the dog has the largest. The sinuses are filled with struts of bone that look something like a hard natural sponge but with larger holes. A puppy, dog or wolf, like a human baby, has a high forehead and domed skull because the brain is large compared to the size of the head, but the adult dog has a raised forehead due to the development of the sinuses as the skull matures. Unlike human frontal sinuses, those in canids are not directly connected to the respiratory system and have no mucous lining, so they do not help in moistening, warming or cooling air during breathing.[12] Until recently their function in canids was not clear, but in the last few years some computer-assisted research has provided a possible answer.

The living dog-like mammal that has the largest, longest sinuses is the spotted hyena. The hyena's sinuses do not stop at the frontals, but continue along both sides of the sagittal crest (the ridge of bone down the center of the skull). J. B. Tanner, et al., used computer simulations of a hyena biting down powerfully with its canines or carnassials, varying the size of or eliminating the sinuses.[13] They found that the frontal sinuses serve to dissipate the force on the lower facial bones during hard biting, spreading it up and back across the frontals. This was also the conclusion of A. Curtis and B. Van Valkenburgh (2014) who found that: "Our results support the hypothesis that frontal

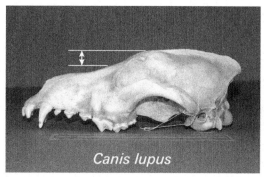 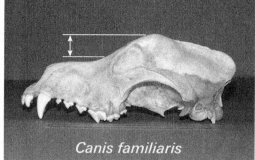

FIGURE 2.2. *The difference in forehead profiles: Wolf is on the left.*

sinuses most often opportunistically fill space that is mechanically unnecessary, and they can facilitate cranial shape changes that reduce stress during feeding." Without a frontal sinus, the force concentrated in areas of the muzzle under the eyes, which in very hard biting created stress high enough to damage the bones. They note that bone-cracking hyaenids and borophagine canids (extinct canids not closely related to any living species) have uniquely vaulted foreheads created by an enlarged frontal sinus, and this vaulting creates an arc of bone that transfers stresses incurred during biting from the facial region caudally along the sagittal crest. So, perhaps the elevated frontals on the dog was part of the ancestor's adaptation to a life style that included chewing on large bones or some other behavior that applied large stress forces to the lower face.

In general, everything else being equal (which it rarely is of course), shorter, wider muzzles have higher bite forces at the canine teeth than long muzzles. This is a simple mechanical advantage related to the distance from the point of attachment at the back of the jaw (the condyle) to the canine teeth, a distance called the "out-lever." Short out-levers generate more power. Long out-levers can close faster. G. J. Slater, et al. *(p. 186)*, found that: "[Small prey catching] canids achieve faster jaw closure by lengthening the jaw out-levers, albeit at the expense of bite force. Short, broad jaws allow large prey specialists to produce large bite forces efficiently by reducing the length of the

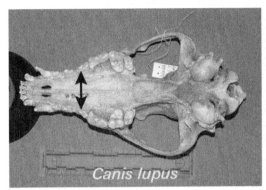 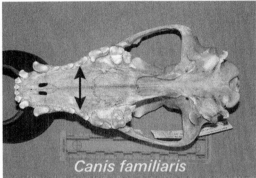

FIGURE 2.3. *The relative widths of the dog and wolf palates. The wolf is on the left.*

jaw out-lever, therefore by increasing the leverage of the jaw musculature. Generalists fall between."[14] Hyenas and cats have very sort muzzles and strong bite forces, and both kill their prey by biting and hanging on. Wolves can have relatively long narrow jaws because working as a pack they kill large prey mainly by slashing and cutting with their canine teeth. Lone wolves rarely take natural prey much larger than themselves (domesticated livestock excepted) because a single wolf does not have the stamina to run down the prey, then hold it at bay, and trying to kill struggling large prey alone with a holding bite could result in injury due to the torque forces on the muzzle which are not dissipated by large sinuses. Other than the young offspring of deer, antelope and gazelles, the smaller canids like jackals generally take prey much smaller than their body size. They have very narrow muzzles and slender teeth designed for fast action rather than hanging on. The generalized, "normal," dog skull and jaw seems to be unique compared to the other *Canis* species.

One of the things about the just-so stories of how the wolf became the dog that does not make good biological sense is the idea that the dog's teeth are relatively smaller, less robust, than the wolf's merely because for thousands of years dogs have not had to use them to hunt, or because humans selected proto-dogs for smaller, less dangerous teeth. Teeth are one of the body parts most resistant to change. The animal's life depends on its teeth being able to do the job they were designed for. In Nature, body parts of a species do not change form unless there is a specific biological reason, some "selective pressure" to change. However, the fact that most dogs have not had to kill their own food since joining humans (the absence of natural selective pressure) is not a good biological reason for change. The teeth would just stay the same because they are not a detriment to the animal's survival. The comparison has not been done, but it would be interesting to know if the teeth of dogs 12,000 years ago were relatively larger than those of recent dogs. Dingoes have dog teeth, not teeth like wolves, even though they have been predators for at least 4,000-12,000 years. The notion that Stone Age humans specifically selected proto-dogs for smaller teeth is absurd. If they allowed wolves to live with them in the first place, they obviously were not worried about the size of the teeth.

My colleague Bonnie Yates and I did the first study comparing dog/dingo teeth and some jaw and skull measurements to those of other canids, replicating the methods used by B. Van Valkenburgh and P. Koeplfi who determined that canids have skulls and teeth adapted to how they make a living, to their lifestyle (ecological niche) as a small, medium or large game specialist.[15] The surprise was dogs do not fit into any of the predetermined ecological categories devised by past researchers. In fact, the dog has a unique set of traits compared to all other canids studied so far. Going by their combination of traits it appears the dog ancestor may have a been adapted to taking fairly large prey single-handedly, or to utilizing more bone in their diet than living *Canis* species. This is discussed in depth in Chapter 6.

Although it has been common knowledge that dog canine teeth are shorter (smaller) than the wolf's, Bonnie and I found that the dog's canine teeth are in fact not shorter compared to skull length but they have a different shape. D. F. Morey was the only previous researcher to actually measure and compare the height of dog and wolf canine teeth (in his study relative to palate length) and found they were equal.[16] Morey did not examine canine shape, but as part of our study we compared the cross-sectional shape of the canine teeth and found that wolf canines are more slender and curve back more. Narrower side-to-side than front-to-back, in cross section wolf canine teeth are oval. Dog canine teeth are more round in cross section. Slender teeth are prone to breaking during the side-to-side torque created by holding on to large struggling prey. Rounder teeth are much more resistant to breaking during a holding bite.[17] A reasonable conclusion would be that the dog, and therefore its ancestor, is adapted to a different hunting style than the wolf.

The bulla, the bone compartment enclosing the inner ear of canids, is an interesting part of the skull anatomy. Very little research has been done on their structure and function. Wolves have what are termed "inflated" bullae, rounded like half of a small egg. Dogs have bullae that are smaller, flatter, with a ridge down the middle (*See Figure 2.4*). The other species of *Canis* have inflated bullae similar to the wolf. We know the bulla serves as a resonating chamber. In his 2001 paper D. V. Ivanov described several carnivore bullae and showed that the internal structure of partitions is similar in related species.[18] Ivanov related their structure to the frequencies of sound the animal can hear the best. Usually the just-so stories ignore this difference in anatomy between dogs and wolves, although a few authors have used the same invalid argument of disuse to explain it: The dog bulla is smaller because dogs do not have to have hearing as acute as the wolf's. Since the acuity of wolf hearing has not been tested, this is just a wild guess. It is a bad guess because apparently the shape and size of the bullae has nothing to do with the level of sound the animal can detect (acuity), and everything to do with the

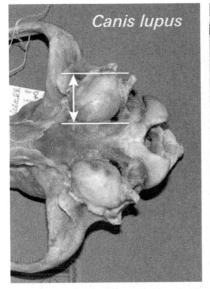
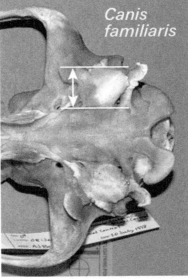

FIGURE 2.4. *Wolf bullae on the left, and dog on the right.*

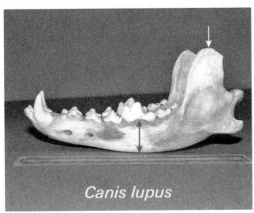
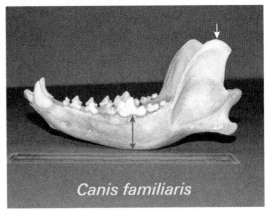

FIGURE 2.5. *The dog mandible is on the right, from the same 2 specimens in previous figures, has only a moderate back-turn at the apex of the ascending ramus. Many dogs, especially small ones, have a more pronounced back-turn.*

frequencies of sounds most biologically important to the species. Bears and raccoons, both omnivores, have bullae very similar in shape to the dog's. But other canids more omnivorous than the wolf have inflated bullae, so the mystery remains to be solved.

The differences in the dog and wolf mandible (lower jaw) have never been examined scientifically or even mentioned in the just-so stories. However, the thicker, deeper area of jaw bone (horizontal ramus) under the dog's carnassial tooth adds strength to the jaw, and would make it more resistant than a wolf's jaw bone to bending in both directions. Like the dog's shorter, wider muzzle, this could be an adaptation for holding onto fairly large struggling prey. The backward curve of the blade of the rear of the jaw bone (ascending ramus) is uncommon among species in the genus *Canis*. It is very rare in Holarctic gray wolves, but present at high frequencies in dogs, Chinese, Arabian, and Indian wolves.[19] Perhaps the smaller southern wolves and the dog inherited this trait from a common ancestral wolf, or it could be a trait that adapts canids to the niche (hunting style) between the jackal and the larger gray wolf. Of course, it also may be a variable trait in all wolves with no adaptive value, increased in frequency in some populations merely by chance. I have been unable to find any explanation of how this shape may be related to function. The only certainty is it is not a paedomorphic trait and was not something Stone Age people could have selected for.

The different angles of adult dog and wolf eyes (angle of the orbits or AO) is another characteristic never examined for their function (and there may be none—the difference could be non-functional), but included in the just-so stories. As wolves were domesticated, dog eyes, one story goes, were selected to be less angled and rounder because humans find this shape less threatening. Puppies, wolf and dog, have more forward-facing eyes with less of an angle because the angles develop as the head matures, the just-so stories say the original dogs were selected to be more appealingly puppy-like. This AO measurement is shown in Figure 2.5. Although not determined that way on a bare skull, it equates to the angle a line through the eyes, corner to corner, would have in relation to a horizontal line across the center of the face. Wolf AOs are usually about 45° and dogs 50° or more. The dingoes and some aboriginal races of dogs have AOs that are intermediate, averaging about 47°. The only wolf that regularly approaches the less acute dog angle is the Arabian wolf.[20] Research has shown that humans around the world have a natural aversion to all eyebrows tilted down to the center at inward angles (\ /) as this is considered threatening or angry.[21] Most graphic depictions of wolves and evil or angry "things" are given eyebrows or eyes with a 45° angle or less (*See Figure 2.6*, which shows the AO of the wolf at about 45 degrees).

FIGURE 2.6. *Wolf Orbital Angle at 45°*

SUMMARY

Is the dog a wolf, the species *Canis lupus*? Does the preponderance of evidence from the skull and jaw indicate they are the same species that should be called by the same scientific name, or a different subspecies, or a different species entirely?

I think the only reasonable answer is today, given the available evidence and based on comparing the gray wolf to the generalized dog and dingo, they are clearly different enough to be considered distinct species. The next Chapter will look at evidence other than the bones of the head that may lead to an answer of the next question: Was the dog ever a wolf?

NOTES

1. Hey, J., 2001; Baker, R. J. and R. D. Bradley, 2006; Wheeler, Q. D. and R. Meier, eds., 2000; Bock, W. J., 2004.

2. Baker, R. J. and R. D. Bradley, 2006; Mayr, E., 1970.

3. Winston, J. E., 1999.

4. The traditional nomenclatural categories above species for the dog, developed after Linnaeus.
   Kingdom: Animalia (the others being Plantae, Fungi, Protista, Archaea/Archaeabacteria, and Bacteria/Eubacteria)
   Phylum: Chordata (animals with spinal cords)
   Class: Mammalia (animals that feed their young with milk)
   Order: Carnivora (mammals that eat flesh)
   Family: Canidae (canids = carnivores that walk on their toes rather than flat-footed like a cat or bear, have a certain kind of skull, and 42 teeth)
   Genus: *Canis*
   Species: *familiaris*

5. Mayr, E., 1970.

5a. Gardner, A. L. and V. Hayssen, 2004.

6. Arnold, M. L., 1997; Monzón, J., R. Kays and D. E. Dykhuizen, 2014; Wheeldon, T. J., et al., 2013; Samonte, I. E., et al., 2007.

7. ICZN, 1957. Opinion 451; ICZN, 2003. Opinion 2027 (Case 3010).

8. Wilson, D. E. and D. M. Reeder, eds., 2005; Vilá et al., 1999; Wayne and Ostrander, 1999; Savolainen et al., 2002.

9. Vilà, C., J. E. Maldonado, and R. K. Wayne, 1999; Vilà, C. and J. A. Leonard,

2012; Ardalan, A., et al., 2011; Brown, S. K., et al., 2011; Sacks, B. N., et al., 2013.

10. Morey, D. F., 1992; Wayne, R. K., 1986; Trut, L. N. et al., 1991; Morey, D. F., 2010; Derr, M., 2011.

11. Drake, A. G., 2011. For true selection effects on the dog skull and brain see: Roberts T., P. McGreevy and M. Valenzuela, 2010.

12. Negus, Sir V., 1958.

13. Tanner, J. B., et al., 2008; Curtis, A. and B. Van Valkenburgh, 2014.

14. Slater, G. J., E. R. Dumont and B. Van Valkenburgh, 2009.

15. Koler-Matznick, J. and B. C. Yates, 2016; Van Valkenburgh, B., 1989; Van Valkenburgh, B. and K. P. Koepfli, 1993.

16. Morey, D.F., 1992.

17. Wroe, S., et al., 2007; Wroe, S., C. McHenry and J. Thomason, 2005.

18. Ivanoff, D. V., 2001.

19. Janssens, L., R. Miller and S. Van Dongen, 2016.

20. Harrison, D. L., 1973.

21. Aronoff, J., B. A. Woike, and L.M. Hyman, 1992.

CHAPTER 3. DOG VS. WOLF

There must be no barriers to freedom of inquiry. There is no place for dogma in science. The scientist is free, and must be free to ask any question, to doubt any assertion, to seek for any evidence, to correct any errors. — J. Robert Oppenheimer. *Life* (10 Oct 1949). Quoted in Lincoln Kinnear Barnett, *Writing on Life* (1951, p. 380).

The differences of the skull and jaw in the wolf and dog are not the only traits that vary between them. The dissimilarity also extends to many other skeletal, physiological and behavioral differences less frequently considered. For instance, in the original 1758 dog description by Linnaeus naming the two species, he used one major, highly visible physical trait to distinguish dogs from wolves. It was an ingenious choice because, other than dogs that have been altered by artificial selection, all dogs and virtually no wolves have this trait. This diagnostic characteristic is the dog's tail.

Historically, scientists used Latin as their common language because it was "dead" and unchanging; Latin words always meant the same thing. Today the universal scientific language is English, although Latin is still used for classification. Linnaeus used Latin for his descriptions and the defining characteristic he gave for the dog vs. the wolf was "*cauda recurva*," meaning "up curved tail." The last half of a normal dog tail curves slightly up, so when the tail is lifted high as a social signal, it looks something like a question mark. This is certainly distinctive as no other canid has this kind of tail, and as usual the just-so stories say this is a product of artificial selection. In fact, the stories are very specific, and say that this high curved tail was favored by the early herders, back when dogs still looked a lot like wolves, so they could easily tell if it was their dogs or hunting wolves out near the flock, thus avoiding sending arrows or spears at the dogs. But, by the time there was livestock to be tended, dogs had already spread around the world. If this explanation is correct, there should be a lot of straight-tailed dogs from places without livestock to tend.

Because the curved tail is practically universal in dogs, the second explanation is based on population genetics (how genes are distributed within and between separate populations). The curved tail is a dominant trait. When combined with the gene (or whatever is inherited that causes tail shape) for a straight tail, the curve will predominate. This second explanation says the mutation for curve happened by chance in only early dogs (proto dog wolves or in the natural species) and this is undoubtedly true because all other living species of *Canis* have straight tails. The dog is unique. There are two alternative stories that can explain when the mutation appeared and how it became universal.

Maybe one of the earliest dogs, when there were only a few of them hanging about a Stone Age camp, had the mutation and for reasons other than the tail shape, that dog's offspring were more successful than others. The curve trait spread in the small early dog population, eventually becoming a "fixed" or universal trait, but not because of the tail. The dogs that had the curve also had some other trait that enhanced survival in the new human-centered niche. It was that unknown trait natural selection favored and the curve was merely carried along with it. Perhaps the gene for straight tail was lost in that early small population when those dogs that had the straight gene were by chance eliminated from the gene pool, a process called genetic drift.

There is a third, less complicated explanation that requires fewer assumptions: It was a natural characteristic before the dog was attached to man. Possibly this was the type of tail the dog's ancestor had, and because it had some useful biological purpose the ancestral dogs that had that kind of tail were, for whatever reason, able to leave more offspring. The up-curved tail is actually an excellent visual signal that can be seen from a considerable distance, which provides enough social advantage to be naturally selected for. The signal value is increased if the tail has a white tip, and if the underside of the tail has longer, paler hair. The New Guinea Dingo [NGD] has the perfect type of tail for sending clear, unambiguous signals to conspecifics, as can be seen in Fig. 3.1.

FIGURE 3.1. *A pair of New Guinea dingoes signaling excitement with their tails.*

This dingo is discussed in Section II, but the important points about them in this context is they are not pack living, normally are highly aggressive to strange canids, and have few signals in their behavioral repertoire that indicate "I am friendly." When one NGD meets another, the tail held high in excitement with friendly intentions is wagged in erratic fashion, and the white tail tip bobs around. If a NGD feels potentially aggressive to another, its tail is up, but either held stiffly still or wagged slowly just slightly side to side. Of course, both tail signals are accompanied by subtler signals of intent, such as ear and mouth positions that help clarify the signal. The two NGDs in the illustration have found potential prey, and their tails are clearly signaling excitement. The bones of the dog's tail are individually straight and no different than the wolf's bones. The curve is a function of the muscles and tendons, and even a fairly curled up tail like a Malamute's can usually straighten out when relaxed.

WOLF AND DOG BODIES

Unlike their skulls and mandibles, the skeletons of wolves and comparably-sized dogs are so alike in general shape that the post-cranial bones are rarely compared in detail. One of the few studies comparing dog and wolf post-cranial elements was by A. Casinos et al. published in 1986. They statistically compared the long bones–the leg bones–of a large sample of 108 dogs of variable types with those of 12 wolves. The largest difference was the wolves had relatively longer and more slender femur (thigh) bones than dogs. The dog exceptions, those with the same proportions as wolves, were a few long-legged breeds: Collie, Ibizan hound, Irish setter, and Scottish deerhound. The most relevant point in the Casinos study was "Because the supposedly primitive breeds, such as Dingoes and Pariahs, have long bone morphology quite different from that of wolves, the wolf condition probably reflects a secondary convergence [toward the wolf] in some specialized dogs." *(p. 78)*

The Casinos conclusion is undoubtedly accurate. Wolves are highly cursorial, meaning they are adapted to chasing fast-running prey for long distances. Their slender body type was designed by natural selection to cover ground fairly quickly and in hot weather to dissipate heat from activity efficiently, which increases the distance they can run. They are, in general, very long-legged and thin-bodied compared to dogs (remember, dog used alone in this discussion means only the primitive types; others are indicated with modifiers). Looking at a wolf in winter coat gives a false impression of body shape. Figure 3.2 is of a wolf in summer coat. The modern dogs that had the most wolf-like legs in the 1986 study are the sight hounds, designed to be highly cursorial through human-assisted selection. Some pariah dogs from hot climates with open plains, such as local races of village dogs of India and Africa, approach the longer-legged sight hound type (see Section II), and these naturally adapted coursers probably have femurs relatively as long as wolves. The specialized sight hounds such as the Saluki were developed from ancestral dog types similar to these natural coursers by at least 5,000 BC, and have continued under harsh selection for function as runners to this day.

FIGURE 3.2. *A North American gray wolf in summer coat.*

It is often opined that dogs have relatively broader chests than wolves, and this is certainly true for many of the breeds specialized for work such as sled pulling, livestock guarding, and retrieving. Dingoes, especially the desert variety, some cursorial aboriginal dogs, and modern sight hounds may have average body widths similar to those of wolves, but no one has done a specific comparison. Dingo leg length and body thickness varies according to the local environment: Desert and plains dingoes are more slender and thinner-bodied, mountain dingoes broader and shorter-legged, and tropical dingoes intermediate between the extremes. The smaller New Guinea dingo is even shorter-legged than mountain Australian dingoes, and as mature adults their chests appear to be relatively wider than those of wolves. The "appear to be" is because no actual measurements have been taken or compared. This is merely the subjective opinion of those who, like myself, have observed several members of both species. The relevant question is, why do the primitive and aboriginal dogs have bodies that are so different than wolves? These populations remained in the same environment and have not been artificially selected for any trait by humans and so have had no reason, natural or artificial, to change their shape or size in at least the last 10,000 years. The answer may be that there has not been any biological reason for the body to change, and so the dog body has not changed very much. Like the skull and jaws, the body still has the form that suited its ancestor to some particular life style.

Another minor difference between the wolf and dog is the apparent relative size of their front feet. The wolf foot looks much broader than the dog's. I started a study to try and figure out exactly how they differ, if the bones in the wolf foot are relatively longer, or if the wolf foot merely spreads out and flexes more due to elastic differences in the tendons. The research, which involves dozens of measurements of each foot bone, is not yet completed. However, after examining the metacarpals (the bones that form the "back" of the human hand and the "pastern" of the dog) of 11 dogs and 34 wolves, it was apparent the difference was not in this bone. The next bone to be measured, the most likely candidate responsible for the larger wolf foot, will be the first phalanx, the bone that forms the top of the paw and the first (longest) bone of the human finger. Figure 3.3 shows the difference between dog (represented by a dingo) and wolf forefeet on a hard surface. I think maybe the wolf, which evolved in northern latitudes, has feet adapted to traveling on snow: Natural snow shoes. The dog has a more compact forefoot similar to those of the majority of canids.

PREHISTORIC DOGS

For those not familiar with the terms archaeologists use to designate the periods of prehistory, called "ages," here is a basic explanation. These cultural "ages," like "species," are rather fuzzy and have somewhat variable meanings and interpretations. Archaeologists have divided human history into approximate ages based on the types of implements or technology the people had during that period, a useful way to organize human history. The period that dogs could first be a possible consideration for

FIGURE 3.3. *Feet of a North American gray wolf on the left and an Australian dingo on the right, illustrating the relative size of the front paws (photos not to same scale).*

association with humans is the Paleolithic, or Old Stone Age. This started with the first recognized use of stone tools some 2.6 million BP, long before modern humans existed, and ended about 10,000 BC. The majority of tools were made of worked stone, bone and antler. It is divided into several smaller periods: Lower (oldest, going back over two million years), Middle, and Upper. The Upper Paleolithic, with well-documented dogs near the end of the period, is considered to be about 50,000–10,000 BC. The Upper Paleolithic is followed by two smaller periods: Paleolithic (50,000–10,000 BC), Mesolithic (intermediate between the other two, 10,000–6,000, but did not occur in all areas) and Neolithic (10,000–2,000 BP). The dates of these periods vary significantly between geographic areas because cultures developed along different time-lines and the designations are based on the types of technology the people had. Paleolithic people were all hunter-gatherers. By the Neolithic, several cultures were starting to domesticate plants and animal species other than the dog, and the first permanent towns appear. After the Neolithic, the timeline is roughly divided into the Bronze Age (3,600–700 BC) and Iron Age (2,000 BC–400 AD). The Historic era starts about 586 BC and of course extends to the present (*See Figure 1.4 in Chapter 1*).[2]

Zooarchaeologists (those who study animals found in archaeological sites) compare the animal bones from Paleolithic sites to those of similar wild species of that time and place, in order to determine if the animals were, or were becoming, domesticated. When there are no skulls, mandibles, or teeth, the size of the long bones (leg bones) is the most frequently used criteria for canid species identification. It is a general truth that when a species is domesticated, at first it becomes reduced in size. Zooarchaeologists who study the domestication of animals hold that size reduction is one of the first signs the animals in archaeological sites, such as aurochs (one of the ancestors of the cow), are becoming domesticated. There have been many reasons proposed for this, such as poor nutrition in early captives giving smaller individuals a better chance to survive, or that it was due to inbreeding.[3]

Since the first agreed-upon domestic dogs are smaller than wolves of the same area and time period, their relatively "reduced" size is considered an artifact of domestication. However, if the dog ancestor was not the wolf, we do not know how much, if any, of the size difference was due to domestication. Without the species specific skull, post-cranial canid bones found in archaeological sites are identified as "possibly dog" through comparison to wild mid-size canid species of that place and time period, such as jackals, coyotes and red foxes. In South Asia and the Middle East the comparison includes the local wolf species, which are smaller than the Holarctic wolf and might be mistaken for dog. Species identification made only from post-cranial bones are frequently considered tentative, because all their size ranges sometimes overlap.[4]

Zooarchaeology became an established science-based discipline only in the 1950s. Until about 75 years ago, after a cursory inventory archaeologists often discarded nonhuman bones or stored them in a museum basement somewhere, and they were never studied in detail. Some of the most "startling" new paleontological discoveries, including some of the oldest dogs, were made when the bones from old excavations, some made as much as 100 years ago, were more carefully examined.

One of the first things that struck me when I started reading about prehistoric dogs was their small size compared to wolves, even the smallest species of extant wolves. The size difference between wolves and the first domestic dogs that everyone agrees on from about 12,000 BC is huge. According to Germonpré et al.'s 2009 paper, the Pleistocene wolves of those times and places where the oldest completely accepted dogs have been found (i.e., Switzerland, France, Portugal, Germany, England, Denmark, Israel, Iraq) are estimated to be about 24–32 inches at the shoulder (estimated shoulder height, or ESH, 60–83 cm). Recent Eurasian wolves have a slightly wider ESH variation, from about 23–35 inches (60–90 cm). The late Paleolithic/Neolithic dogs of 15,000–9,000 BC were in general between ESH 12–18 in. (30–45 cm), although a few have an ESH of about 23 inches (60 cm). Later, in the Neolithic and Bronze ages some dogs were even less than 12 inches (30 cm) ESH. While they will not be considered further as they obviously were domesticated well before migrating into North America with humans, and so have nothing to tell us about the origin of the dog, according to B. E. Worthington (2008), American prehistoric dogs from 7,450–3,370 BC had ESHs of about 13–19 in. (33.24 cm to 48.69 cm), and so were very similar in size to the Old World early dogs they descended from.[5] Figure 3.4 is an Alaskan wolf compared to an Australian dingo. The wolf is imagined to be about 28 in. (71 cm) at the shoulders, average overall for Holarctic gray wolves with sexes combined. The Australian dingo's height is pictured as average for dingoes, about 20 in. (50.8 cm). This dingo would be on the large size for some Neolithic dogs, but around average for all prehistoric dogs. Here it is serving as a stand-in for the ancestral dog, whether that is considered to be the earliest point at

FIGURE 3.4. *The relative size of an average wolf and Australian dingo.*

which the wolf had become the dog, or the dog formed as a natural species.

Undoubtedly, if an individual tamed wolf had to make a living scavenging meat, bone, and vegetable scraps from gatherer-hunters, it would be malnourished and therefore stunted in development. If, in spite of that, it was able to reproduce, those offspring that were genetically smaller would have a better chance to survive the poor diet. Over many generations, the average size of the population would go down naturally, but being reduced to half size would be unusual in a relatively short period of time. It seems, then, that to have small dogs by about 12,000 BC either the domestication of the wolf started much earlier than the 15,000 BC often estimated from genetics and the archaeological record, or the original dog was a smaller type of wolf than the typical gray wolf. The oldest estimated date of wolf and dog separation from DNA dating is about 36,000 years, with a bottle neck (greatly reduced number of individuals) in dogs at about 27,000 BC.[6] If the wolf is the direct ancestor of the dog, the 36,000 BC date would be the most likely, as natural size reduction of that magnitude could easily have taken place in 10,000 years.

Recently, an interesting series of papers appeared about some much older large "wolf-like dogs." These canids were described from skulls found at mammoth hunter sites in Siberia, Předmostí in the Czech Republic, and a cave in Belgium.[7] Their size was startling since they are much larger than the largest agreed-upon dogs of about 14,000 BC. These Paleolithic canids are dated to about 33,000 BC and their ESH is 24–32 in. (61–83 cm), essentially the same as wolves. There are very similar large "Paleolithic dogs" from Russian (Eliseevichi I) and Ukrainen (Mezin) mammoth hunter sites, but those are about 18,000 years younger (15,000 BC). DNA studies done on the Czech and Siberian specimens did not reveal any mtDNA haplotypes (this means a combination of DNA sequences that are near each other on the chromosome and so are transmitted together to offspring) that have been found in living dog or wolf populations, and the authors concluded that these races of "dog" must have died out. But there was no reason offered for this extinction of a supposed human-associated genetic line of "dog." It could be that genes inherited from these canids are present somewhere today in the millions of untested dogs and wolves. As discussed in Chapter 1, these large wolf-like canids were determined to be dogs mainly because they have very short muzzles (even for average dogs) and wider craniums compared to typical wolves. Figure 3.5 is a 33,000 year old Paleolithic "dog" skull from Altai, Siberia.

Some of these Paleolithic dog-like canids have crowded teeth with no spaces between them, or teeth that have to sit at an angle to fit into the jaws, a sign of recent muzzle shortening which the teeth have not yet adapted to. The tooth crowding was assumed to be a result of rapid size reduction during domestication. When they lived, it was a period of recurring ice ages and there were abundant very large animals including bison, musk oxen, and mammoths on the plains below the glaciers and after the glaciers retreated. Today much of that area is conifer forest. There were also many species of moderately large prey like red deer and horses. In addition to wolves, there was a wide array of larger predators: Ancestors of the spotted hyenas, multiple species of bear, and large cats, including lions and tigers.

While it is possible the Paleolithic "wolf-dogs" were partial ancestors of at least some of today's dogs, and their mtDNA either has not survived or remains to be found in the millions of dogs not sampled for the DNA studies, it is much more likely they were a species or subspecies of wolf adapted to a different hunting style than the more common *Canis lupus* wolves. Perhaps they evolved in an area without hyenas, and so could develop the same hyena-like life style of scavenging carcasses, cracking bones, and killing large prey by hanging on. Chapter 2 discussed the differences between the wolf and dog skull and jaw: Dogs have a shorter, wider muzzle and more rounded skull with a heavier jaw bone than *lupus* wolves, and these short-faced wolves appear to have similar adaptations, but are larger than the earliest agreed-upon dogs. Regular *lupus* wolf remains are much more common at the mammoth hunter sites, perhaps an indication there were fewer short-faced wolves, or, if humans killed the wolves in defense of their kills (or in defense of what they were scavenging), the short-faced wolves may have hunted/scavenged singly or in pairs, rather than packs, so they were less frequently killed at the site. Another alternative is the short-faced wolves were not killed in greater numbers because they did not compete directly with humans for carcasses, arriving only after the people had abandoned carcasses.

It is reasonable to speculate that there were two co-existing large wolves when there was a super abundance of large prey species. The short-faced specialization would have been a way to avoid direct competition with *lupus* wolves. The *lupus* species, as today, was specialized for group hunting by slashing prey to disable it. The short-faced species perhaps specialized in killing large prey by hanging on and scavenging megafauna carcasses, going extinct along with the largest prey species. Such specialization often

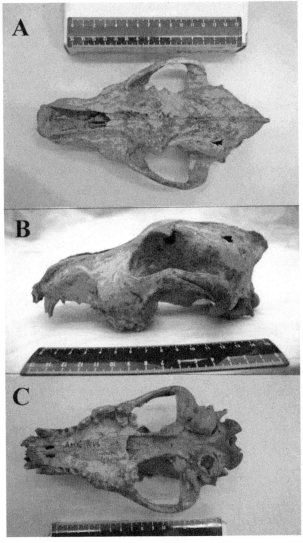

FIGURE 3.5. *The canid skull from Altai. Note the raised forehead, very short muzzle, and robust zygomatic arches (cheekbones), all likely adaptations to killing larger prey with a holding bite or to bone cracking.*

develops when there is a high degree of competition from similar carnivores. Radio isotope analysis of the bones of the lupus wolves and short-faced wolves present at the same site show that the lupus wolves were eating mostly the largest prey, mammoth and bison, and the short-faced wolves ate mostly reindeer and horse.[8]

There was a comparable short-faced wolf in eastern Beringia (Alaska), radiocarbon dated to be from 12,500 BC to beyond the radiocarbon dating limit (about 50,000 BC). Since there were 56 Beringian wolves to examine, the sample size was much larger than the handful Paleolithic short-faced wolves. The Beringian wolf had a similar short, broad muzzle and its mtDNA also has not been found in modern wolves or dogs. Leonard et al. (2007, p. 1146) say: "Moreover, skull shape, tooth wear, and isotopic data suggest that eastern-Beringian wolves were specialized hunters and scavengers of extinct megafauna. Thus, a previously unrecognized, uniquely adapted, and genetically distinct wolf ecomorph suffered extinction in the Late Pleistocene, along with other megafauna." It seems this wolf species was widespread over all of Beringia and Eurasia during the late Pleistocene, and at least in Beringia was present before modern humans arrived. D. Morey, a zooarchaeologist, questioned the identification of the Paleolithic wolf-dogs as dog, cautioning against conclusions based on such a small sample size, and pointing out flaws in the studies by Germonpré and colleagues. M. Germonpré et al. answered his criticism upholding their conclusions.[8]

The identification of two of the wolf-dogs as short-faced wolves, not proto dogs, was later supported in a 2015 paper by A. G. Drake, M. Coquerelle and G. Colombeau.

Their 3D morphometric study used computers to scan and measure the skulls and analyze them in three dimensions. The Goyet (Belgium) skull they studied is dated to 31,680 +/− 250 BC and the Eliseevichi (central Russian plain) skull, identified by the designation MAE447/5298, was dated to 13,905 +/− 55 BC. Both were first identified as Paleolithic dogs. They compared these skulls to both ancient and modern dogs with generalized skulls, and to prehistoric and modern wolves.[10] Their conclusion was:

> We demonstrate that these Paleolithic canids are definitively wolves and not dogs. Compared to mesaticephalic (wolf-like breeds) dog skulls, Goyet and Eliseevichi MAE do not have cranial flexion, and the dorsal surface of their muzzles has no concavity near the orbits. Morphologically, these early fossil canids resemble wolves, and thus no longer support the establishment of dog domestication in the Paleolithic. These skulls did not have dog-like raised foreheads, so did not have what dog people call a "stop" just above the eyes and did not have the cranial flexion mentioned before as a dog trait. *(p. 1)*

Then again, these short-faced wolves could have been a second natural species of dog rather than of wolf, although they were much larger than recognized dogs from later times in East Asia and Europe. Maybe, just maybe, there was more than one species of dog with similar specializations, separated by space and time. The large short-faced "wolves" could have evolved on the plains with the *lupus* wolves from the same ancestor as the dog, while the smaller southern dog species evolved in forests (discussed in Chapter 6). But, in any case, the short-face wolves probably were not wolves on their way to domestication, as there was little change in them over a span of at least 18,000 years. In that length of time if they were wolves on the way to becoming dogs, they would have become smaller.

DOG BRAIN TRAITS

The domesticated populations of most mammal species have relatively smaller brains than their ancestors, which is apparently a side-effect of domestication. The dog has a smaller brain in relation to body size than the wolf, which is also assumed to be a result of domestication. Older estimates by W. Herre (1970) and others concluded that the dog brain is about 10–30% relatively smaller than that of the wolf. All kinds of speculation abounded about the reasons for this. Dogs, they said, have not needed to be as intelligent as wolves—this is another invalid argument from disuse, like the tooth changes. The size of a brain is not correlated with degree of intelligence. Einstein, for instance, did not have a larger brain than an average person. Or, they speculated that dogs could depend on humans for protection and food, so the parts of their brains devoted to their senses were reduced. However, there are no neurological or anatomical studies that show a direct relationship between the size of various areas of the brain and the acuity of any sense. Some brain traits, such as the size of the neocortex (the outermost layer of the brain) in mammals, have been associated with increased cognitive ability and the flexibility of behavioral responses ("thinking" and changing behavior by choice rather than being limited to inborn behavior responses), but I could find no direct comparison of the neocortex size of the wolf and dog. Rodents have a rather smooth neocortex, and humans and other larger mammals — like wolves and dogs — have folds (sulci) and ridges (gyri) that allow a greatly expanded neocortex surface area without much increase in brain volume. Brains are, after all, constrained by skull anatomy, which is adapted to the species' particular niche, and by body size.

D. Kruska studied the effects of domestication on the brains of many mammals.[11] He found that the hippocampus and limbic system parts of the brains in most domesticated species are significantly smaller than in their wild ancestors. These systems are involved in emotions and memory, and therefore selection for less reactive or tame behaviors may be responsible. Brain architecture in wild species is somewhat correlated with their mode of life. For example, the brain of whales, adapted to an aquatic life, have very small olfactory bulbs. Both progression and regression of the size of sense-organ-related brain parts occur during evolution of adaptations to different environments and niches. Apparently domestication, which selects for ease of handling and acceptance of confinement/control by humans in a very specialized environment, results in changed brain structure and size in species as different as rats, pigs and sheep.

As usual, the dog was mainly compared to the Holarctic wolf. Assessed against that wolf, the dog does have a relatively smaller hippocampus and limbic system and slightly smaller olfactory lobes. It would be interesting to see how the dog compares in limbic system development with the South American bush dog, as they have been reported to easily become very tame in captivity and friendly toward people, much "calmer" (less reactive) than tamed wolves. [12]

In an almost forgotten 1975 study, V-H. Hemmer compared dingo skulls to those of the Arabian and Indian wolves, which are close to Australian dingoes in body size. These wolves have teeth that are less robust than the Holarctic northern wolves, because typically they hunt only moderate size prey such as small mammals and deer. Recent genetic studies of Indian wolves (Arabian wolves, the other small southern wolf, were not included in the study) by R. K. Aggarwal, J. Ramadevi, and L. Singh (2003) indicated they have been a separate evolutionary line from the Holarctic wolf for more than 100,000 years. While Indian wolves are not as genetically close to dogs as the Holarctic wolf, indicating that wolf is not the dog ancestor, their skulls and teeth show strong similarities. Thalman et al. (2013) note that Australian dingo and Indian wolf skulls are alike, except that the Indian wolf has the typical large, rounded wolf bulla, and the dingo has a bit higher forehead. Hemmer plotted brain to body weight for dingoes and wolves and found that, compared to Indian and Arabian wolves, dingoes have the same relative brain size. Hemmer's conclusion was that the dog was probably domesticated from a canid that already had a relatively smaller brain. He said (English summary p. 98):

> The teeth of primitive dogs (studies on dingo skulls) can only be traced back to a rather unspecialized stage in the evolution of wolf teeth such as is represented up till the present in South-west Asian wolves (*Canis lupus pallipes, Canis lupus arabs*). Small relative brain sizes, which extend directly into the relevant scatter area of dingoes, are to be found in just these wolves in contrast to those of the northern Palaeaearctic [Holarctic]. Consequently, the theory that the relative brain size was considerably reduced (10–30%) on domestication, as advocated especially by Herre and his collaborators, is entirely without support for the origin of the dog. An alternative to be discussed is that domestication of a wild species took effect in each case with such forms as have the smallest relative brain size for the species in question.

The brains of dogs and wolves differ in other ways. Few scientists, except Hemmer, Kruska (1988), and Herre (1970) (with their collaborators), have ever systematically compared the anatomy of species-specific canid brains. D. L. Atkins and L. S. Dillon (1971) studied those of the dog, gray wolf, red wolf, coyote, black-backed jackal, and golden jackal. They state that the genus *Canis* is "noted for conservative evolution of the anatomical characters customarily used in mammalian taxonomy." *(p. 96)*, and this often makes determining relationships more difficult in this species group (they published the study before canid genetics were available). The anatomy of a species' brain is apparently very stable. As long as they have not been artificially selected for mutations like very short muzzles, the dog's brain stays the same even when the skulls are extremely modified, as in giant and miniature dogs. Therefore the size, shape, and complexity of the cerebrum (the higher brain: The "thinking" and integrating part that includes the neocortex) and the pattern of cerebellum (the lower and most ancient part, mostly devoted to automatic functions) anatomy can be used to tell something about whether or not an individual belongs to a given species. The brain alone is not enough to establish certainty. Other traits must be included in the analysis. Atkins and Dillon (1971) concluded there are two main groups in the *Canis* species: One includes only the red and gray wolves, and the other comprises the remaining species examined. They state *(p. 101)*:

> The coyote, domestic dog, and both jackals compose a second group in the mutual expression of some different variation of these traits. In several instances, jackals or the coyote (or both) share certain features with wolves, but in no case does the cerebellum of *C. familiaris* display wolf like characters without at least the golden jackal or coyote (and occasionally both) showing the same feature.

Their conclusion was *(p. 104)*:

> The cerebella of the coyote, domestic dog and the two species of jackal were described above as being more closely allied than any of these are with the wolves. This implies that the ancestry of the dog is probably common with one or all of the jackal species rather than immediately from the gray wolf ... Indeed, wolves show numerous traits distinct from the other species considered, a relationship that agrees with serological and biochemical comparisons.

Since the distant ancestors of the jackals and wolves went to Eurasia from North America over the Bering strait some time about 2 million years ago, and they were also the ancestor of the coyote, this is not as odd a conclusion as it seems. Apparently, the dog retained some of the features of the common ancestor, and the wolf brain developed changes in those traits.

The little comparative research available indicates the dog cerebellum (the most primitive part at the lower rear of the brain that controls movement coordination) is most similar to those of the jackals and coyote rather than the wolf, although none of the species were identical. Many scientists dismiss the Atkins and Dillon study saying the cerebellum is not a good character for determining the species. Maybe they are right. But, perhaps they believe the dog to be a domesticated wolf. If the results show the wolf and dog are different, then their conclusion "must be" the cerebellum is too variable within a species to be an important characteristic. If the cerebellum within a species is as different as the wolf and dog samples Atkins and Dillon examined, it could not be used to determine species. In that case, that part of the brain would be termed "nondiagnostic." The problem is each trait they scored was a bit variable (like all of Nature) and their sample size was small, not enough to rule out the chance variation factor. Hopefully someday another researcher will become curious and expand the number of cerebellums compared, finally proving or disproving the dog and wolf differences.

In his 2010 book on dog domestication, D. Morey discounts the importance of the differences in the cerebellum, citing a study by U. Will (1973) in which she said this area of the brain is not of "taxonomic significance" in the genus *Canis*. Morey proposed, based on a comparative cerebral [neo]cortex anatomy study by G. Lyras and A. A. E. Van der Geer (2003), that there were no significant differences worth mentioning between wolves and dogs. The cerebrum is the larger, main part of the brain that controls all voluntary actions. Lyras and Van der Geer (2003) found large differences in the form of the cerebrum that divided the canids into two main groups, the dog-like and the fox-like. Morey *(p. 196)* said "For present purposes, it is useful to emphasize that brain structures can change, accompanied by little or no discernible genetic change. The case of dogs versus wolves exemplifies that point well." But, even in the cerebrum there is at least one difference between dogs and three species of wolf: The absence of one of the sulci (furrows). Lyras and Van der Geer *(p. 513)* say: "*C. simensis, C. lupus, C. rufus*, and the two examined specimens of *Dusicyon australis*, have three sulci in that region: The proreal sulcus, the intraorbital sulcus and a third sulcus, which is not found in the domestic dog, forming the dorsal boundary of the anterior portion of the orbital gyrus." The Ethiopian wolf has no fossil record of ancestry, but its closest genetic relatives are the wolves. *D. australis* is the extinct Falkland Island wolf, which was distantly related to the South American maned wolf *(C. brachyurus)* and so not a "true" wolf but more fox-like. Its direct ancestors have never been located on the mainland, and it is a mystery when and how it got to the islands. The other species that, like the dog, have only two rather than three sulci are the coyote, dhole, African hunting dog, and gray fox.

Morey, in a footnote, acknowledges there are differences: "Again, dogs differ from wolves in brain structure, even as the two animals are genetically nearly identical. Quite simply, in canids at least, the details of the cerebellar structure do not speak to ancestor-descendant relationships." *(p. 196)*. While surface brain anatomy may in general not be a good indicator of "ancestor-descendant" relationships for species of canids that diverged from common ancestors hundreds of thousands of years ago, it seems unlikely it would be true of two evolutionary lines–dog and wolf–that supposedly separated at most only about 40,000 years ago. So, the fact that some brain traits of dogs do not resemble wolves' must mean something, but exactly what is currently a mystery.

THE AMYLASE GENE

Recently, geneticists looking for areas of the dog genome that might have changed because of domestication found that many dogs have more copies of one gene that codes for amylase, the enzyme that helps digest starch. E. Axelsson et al. (2013) sampled 60 dogs from 14 breeds and 12 wolves from diverse populations. In addition to the increased copy number of the amylase gene, they tested for the expression of the gene in tissues. They found that dogs did have increased ability to absorb glucose (the product of starch digestion) compared to wolves. Several other genes related to starch and fat digestion, regulation of anatomical and neurological developmental processes, and sperm-egg recognition (binding) are different in the dog compared to the wolf. Their conclusion was *(p. 362)*: "We propose that genetic variants within these [amylase] genes may have been selected to aid adaptation from a mainly carnivorous diet to a more starch rich diet during dog domestication." The physiological functions (results) of the other gene differences are unknown. More recent studies (personal communication with researchers) looking at the amylase gene have found low copy numbers (average under 5) in wolves, Australian dingoes, some Arctic landrace breeds, compared to extremely variable copy numbers in modern dogs, from one to over 25 copies. They tested physiological amylase levels, and high gene copy numbers did relate to increased amylase expression. It is certainly true that most village dogs around the world have survived on a carbohydrate based diet for millennia, so better utilization of starch would be beneficial for them. Until other canid species are sequenced for the amylase gene, we will not know what the normal variation is in canids. Perhaps the dog was "pre-adapted" to utilizing starch merely because the dog ancestor had a more variable amylase copy number, and when humans developed agriculture about 10,000 years ago, some thrived on the left-overs, a selective advantage.

THE DOG'S WHITE MARKINGS

In addition to the different shapes mentioned previously, wolf and dog tails have another dissimilarity: color. Wolves always have self-colored or dark tail tips. The dingoes and aboriginal indigenous dogs world-wide can have self-colored tips but have a high frequency of white tail tips. Because it is so ubiquitous in dogs that have not been under artificial selection, this white tip must have been either an ancestral trait, or appeared and then became common in the first population of early human-associated dogs. The dog tail tip is an exception because all the species in the *Canis* genus, except the dog, have dark or self-colored tips. However, a few more distantly related canids like the red fox, maned wolf, and extinct Falkland Island wolf have white tail tips.

The white tail tip is associated with another trait of dogs that has been considered a product of human selection, white markings on feet and chest, sometimes with a white face blaze and neck spot. Puppies of many canid species, including wolves, may have white markings on feet, front of the neck, chest, and on the belly. The lack of color in true white areas (as opposed to very pale cream) is caused by the melanocytes failing to migrate all the way into that area. In the case of small white areas in normally solid color animals, melanocytes—the skin cells that make color pigments—fail to make it all the way from their origin in the embryonic neural crest (a stem cell origin area on the dorsal or spinal side of the embryo from which many cells types are developed) to the most distant body areas. These white marking usually disappear because melanocytes continue to migrate a bit through the skin as the animals mature. Sometimes they do not disappear, and are sometimes seen in adult wolves, and in mammals as different as foxes, bears, gophers, and squirrels. Some solid color dogs also have these non-genetic white markings.

Without getting involved in complicated color genetics, there are also at least two types of inherited white markings in dogs. Each type seems to have "modifiers" that decrease or increase the amount of white *(See Figure 3.6)*. The most common pattern of white in dogs is markings on the front of the face (a blaze), neck (just some spots or whole neck), chest, belly, feet (sometimes high enough up the leg to be called "stockings"), and the tail tip. This pattern is called "Irish white" markings after the strain of laboratory rat it was first described from. This is the type that some dingoes and all of the captive NGDs have. The two pictures available of wild NGDs show white on the chest and the one that had the feet visible, on the toes. No one knows if the wild NGDs have more extensive white markings. The pattern probably has become increased in size in captive NGDs due to modifiers accented during extensive inbreeding in captivity. However, pure Australian dingoes often have white around their nose pad, white patches on the back of their necks, and white stockings, so they too carry the Irish gene.

The third type of white marking is commonly called "piebald." In addition to the Irish markings, piebald results in areas of white spotting on the main body, from just a "splash" to mostly white with spots of color. The extension to "extreme piebald" results in an almost all-white dog, and is frequently accompanied by deafness or eye anomalies (melanocytes play a vital role in hearing and eye development). While the ancient village and pariah dogs may have all three types, the dingoes and some ancient races like the indigenous Thai dog are never piebald.

As in the dog, the frequent white markings on the neck and feet in wild adult red foxes is probably due to some modifier connected to the presence of the white tail tip that extends the markings beyond the usual non-genetic type discussed above. These modifiers were enhanced in the famous Russian tame-bred foxes, resulting in a small percentage of white face blazes. In other fox breeding programs they are now producing the piebald pattern of spots on a white background found in dogs.

If the dog is not a wolf descendant, it is possible that the ancestor had a white tail

FIGURE 3.6.
These three Thai ab-original dogs illustrate all three coat patterns for solid color dogs. From the top: Unmarked, assumed non-genetic chest and foot pattern, and Irish white.

tip. Perhaps the extension to Irish markings just happens occasionally when the white tip is selected for in nature to enhance signal visibility. The incidence of white markings increases significantly in certain solid-color animals living in close proximity to humans, probably because these populations are made of up individuals that are highly resistant to stress (better at coping with anthropocentric habitats and disturbance). Many of the pocket gophers from the area around University of California, Berkeley had white markings, while those from uninhabited areas did not. City flocks of blackbirds often contain individuals with white feathers, while country flocks rarely do.[13] The production of melanin (the colored pigments in mammals and some birds, either red phaeomelanin or black eumelanin) is tied to neurotransmitter and thyroxin production because all are based on a protein called tyrosine. Tyrosine is involved in the production and activity of the neurotransmitters serotonin, GABA (Gamma-aminobutyric acid), epinephrine (adrenalin), and norepinephrine (related chemically to adrenalin) in the brain. A small minority of the genetically tame Russian experimental foxes developed Irish and piebald spotting, and all the tame foxes had significantly higher levels of serotonin than normal foxes. The human-friendly foxes are not "dog-like" in their behaviors and in fact the rigorous selection for tameness made them even more resistant to the negative effects of social and physical isolation than average dogs.[14] Higher serotonin levels are implicated in reduced (from normal levels) aggression and, along with adrenalin and norepinephrine, are activated in the fight-or-flight responses of many animals. Thus, reduced melanocytes seem to be connected to resistance to stress and reduced fear. Maybe compared to the wolf, the dog ancestor had a neuroendocrine system already adapted to withstand greater environmental stress, with comparatively elevated serotonin and reduced emotional reactivity due to lower adrenaline. This was then accentuated in the population that adjusted to living around people and could be the underlying reason primitive dogs have white markings in addition to the tail tip, like the Russian tame foxes.

NEUROTRANSMITTER DIFFERENCES

In 2004 P. Saetre et al. published the results of their study comparing the neural transmitters of dogs and wolves. They analyzed the gene products (neuropeptides and messenger RNA) in various parts of the brains of wolves, dogs and coyotes. They found that two neuropeptides (abbreviated CALCB and NPY) active in the hypothalamus (a brain area responsible for several metabolic activities of the autonomic nervous system including making pituitary releasing hormones) were at the same level in coyotes and wolves, and very different in dogs. They concluded *(p. 205)* that:

> Our results suggest that strong selection on dogs for behavior during domestication may have resulted in modifications of mRNA expression patterns

in a few hypothalamic genes with multiple functions. This study indicates that rapid changes in brain gene expression may not be exclusive to the development of human brains. Instead, they may provide a common mechanism for rapid adaptive changes during speciation, particularly in cases that present strong selective pressures on behavioral characters.

Comparative studies of wolf and dog neurotransmitter levels are difficult because they require fresh brains to test, and wolf brains are not easily available. At least these researchers mentioned differences in neuropeptides may be related to adaptive changes that involve behavior and did not limit this to domesticated animals.

In an interesting study published in 2012, F. W. Albert et al. looked for gene expression differences in the prefrontal cortex of the brain in three pairs of domesticated and wild species (dogs and wolves, pigs and wild boars, and domesticated and wild rabbits), some of which could correlate with behavioral differences. They found only 30 genes (0.2%) that had different expression between dogs and wolves, 31 (also about 0.2%) between domestic and wild rabbits and 75 (0.5%) between pigs and wild boars. The function of these genes varied widely, and the changes were not consistent among the domesticated varieties, indicating that the majority of expression differences are unique to each domestication event. This suggests that behavioral domestication has taken place through different genetic routes in different species. The gene that differed the most between dog and wolf was transketolase-like 1 (TKTL1) which had about a 47-fold higher expression in dogs than wolves This gene codes for an enzyme that promotes anaerobic (does not use oxygen) glycolosis, an inefficient but fast type of ATP production (the energy molecules that cells use) that is prominent in cancer cells and during physical exertion that temporarily uses up all the oxygen available to muscle cells. The gene in wolves that was the most different from dogs was AP5Z1/KIAA0415, which may be a subunit of a protein that transports other proteins in cells. Overall, domestication seemed to be responsible for only about 4.3% of the difference in gene expression between wild and domestic varieties. The significance of these findings is that very small changes in regulation of a few genes can cause fairly large differences in behavior.

BREEDING SEASONS

There are at least two characteristics of modern domestic dogs that are unquestionably due to domestication effects. Wolves have one breeding season a year: Spring in the Holarctic and fall (after the rainy season) in South Asia. Dogs can have multiple breeding seasons (but usually no more than two) in 12 months and some larger breeds cycle longer, between 11–18 months apart. This means that most dog estrus cycles are not dependent upon environmental factors. Dogs can cycle any time of the year. This nonseasonal repeat cycling is often given as a major difference between dogs and wolves, a "domestication effect," at least by those not familiar with aboriginal dogs and dingoes. The explanation given for the difference is that, after domestication, the dog's physiology no longer had to be tied to the environment because humans provided shelter and food year round, and the reduced natural selection allowed the breeding season to vary. Another possible cause, brought to light by the Russian tame fox experiment, was that some (a small minority) of the genetically tame foxes had "odd" estrus cycles either at the wrong time for foxes (normal is late winter) or more than once in 12 months, another physiological change perhaps attributable to strong selection for tameness.

The problem with referring nonseasonal breeding cycles to a domestication effect is apparent in dingoes and many aboriginal races of dog (the ancient free-ranging, naturally breeding village dogs: see Part II) that do have seasonal breeding seasons. Dog estrus cycles not entrained by environmental cues must have originated long after full domestication. For example, New Guinea dingoes, African Basenjis, Indian native dogs, Malayan village dogs, and Thai village dogs have one breeding season a year, in late July and August, with most of their pups born in October–November, after the wet season. The Australian dingo breeding season varies by latitude and local environment, but they also only cycle once a year. This is a strong indication that the original dog was not derived from a Holarctic wolf. If it had been, the most ancient and untampered-with races of dog would have a late winter breeding season like that wolf. We know the Indian wolf, which also has a late summer reproductive season, is not ancestral to dogs because the genetic distance between them is too great. Multiple breeding seasons may have also been directly or indirectly selected for in modern dogs by people favoring high producing females, because this gives the owners a better chance to get the right puppy for a specific purpose, such as herding, and to have excess puppies to trade, sell, or, in some cultures, to eat. Multiple seasons, or seasons out of sync with the environment, are the result of domestication and artificial selection, but are likely based on a natural annual estrus timed almost six months away from the Holarctic wolf season.[15]

COPROPHAGY

The last difference between dogs and wolves that I think may be a species-specific trait is a behavioral difference that most people find unpleasant to discuss. This is coprophagy, which means consuming feces. Many animals eat feces, from their own or other species, because feces often contain a lot of nutrients that were not absorbed. Pigs and dogs frequently eat feces of herbivores, predators and humans. In fact, in many rural villages in nonindustrial societies that lack sanitation systems, dogs and pigs are extremely important as "sanitary engineers." They keep garbage and feces in latrines cleaned up. In modern industrialized society people are disgusted if their dogs eat feces, especially their own or other dogs' (eating herbivore feces, like from horses or deer, is apparently less objectionable to many and "expected"). In fact, coprophagy is often listed as an abnormal behavior in dogs, and speculation abounds about why some dogs do this. Maybe, they say, the dog lacks some enzyme or vitamin, or maybe it was locked up with its own feces so long sometime during its life it just started eating it out of boredom and developed a habit. To deter this behavior, chemicals that turn feces bitter are fed to the dogs, or diets are manipulated to try and supply whatever the dog is lacking so it will stop. Some recalcitrant dogs have even been subjected to shocks from electronic collars. The latter usually works, because the punishment is severe enough and happens only when the dog is near feces, and the dog may even become fearful of feces.[16]

Wolves, as far as I could discover, do not eat feces. Searching the literature on both captive and wild wolves did not turn up any mention of this, so I asked two people with extensive experience with captive wolves about it. If this behavior is present in wolves, captive wolves would be the most likely to express it because they are confined in areas where feces are commonly available and have fewer things to occupy their time. Pat Goodman of Wolf Park and Lori Schmidt of the International Wolf Center, together representing about five decades of caring for and observing dozens of wolves from birth to death, both reported never having seen a wolf eating feces. So, while we cannot say no wolf ever engaged in coprophagy, it seems it could at best be a rare behavior and not typical of the species. The fact that dogs are notorious for eating feces, and wolves do not indulge, may seem to be just an unimportant tangent to the question: Was the dog ever a wolf. However, in Chapter 6 it becomes one of the most important factors.

SUMMARY

Dogs are different from wolves in many ways, all of which have been, without serious consideration, assumed to be the results of changes brought about by domestication. Some of these assumed effects are not even valid differences. For example, in their 2013 paper about the amylase gene duplication in the dog genome, E. Axelsson et al. toss out this statement in the second paragraph about known domestication effects in the dog: "Dogs also differ morphologically from wolves, showing reduced skull, teeth and brain sizes." *(p. 360)*. Actually, direct comparisons have shown the dog skull and teeth are not relatively smaller than the wolf's, merely different, and that the dog brain is relatively the same size as other medium-size canids such as the Indian wolf and jackal. Several of the unique dog traits could have been adaptations of the dog ancestor, a different "wolf" than the one the dog is always compared to, the Holarctic *Canis lupus*.

NOTES

1. Trut, L. M., 1999; Trut, L., I. Oskina and A. Kharlamova, 2009.
2. Butzer, K. W., 1982; James, H. V. A. and M. D. Petraglia, 2005; Lee, R. B. and R. H. Daly, eds., 1999; Straus, L. G. et al., eds., 1996; Wolpoff, M. H., 1999.
3. Brothwell, D. R. and E. Higgs, eds., 1969; Gifford-Gonzalez, D., 1991; Grayson, D. K., 2014.
4. Although many early dog bones were found in burials, most archaeological bones come from what are called "middens," the garbage piles of the distant past. Human made middens contain the bones or shells of animals eaten (skin, hair, and plant material are rarely preserved), along with archaeological artifacts. The bones survive because they get rapidly covered by subsequent garbage (were not exposed on open ground) and there are so many of them, a least some survive. They know the animals were used as human food because of the cut and burn marks on the bones, and the type of breaks. Animal bones can reveal quite a lot about the life of the people and what the local environment was like at the time. For instance, the animal species present tell us if the site was likely to have been in forest, mixed woodland, or open grassland because many species specialize on types of plants available in specific habitats. Pollen, seeds, and other vegetation remains can provide more detail, including the probable average summer and winter temperatures and rainfall. But, before about 50 years ago, pollen and micro evidence was not even collected or examined.
5. Burleigh, R. et al., 1977; Clark, K. M., 2000; Clutton-Brock, J. and N. Noe-Nygaard, 1990; Colton, H. S., 1970; Dayan, T., 1994; Gonzalez, T., 2012; Haag, W. G., 1948; Harcourt, R. A., 1974; Lawrence, B., 1967; Morey, D. F., 1992; Olsen, S. J., 1985; Pionnier-Capitan, M. et al, 2011.
6. Ardalan, A. et al., 2011; Deguilloux, M. F. et al., 2009; Ding, Z-L. et al., 2012; Oskarsson, M. C. R. et al., 2011; Pang, J.-F. et al., 2009; Savolainen P. et al., 2004; Verginelli, F. et al., 2005.
7. Crockford, S. J. and Y. V. Kuzmin, 2012; Germonpré, M. et al., 2009; Germonpré, M., M. Laznickova-Galetova and M. V. Sablin, 2012; Morey, D. F., 2014.
8. Germonpré, M. et al., 2009.
9. Germonpré, M. et al., 2015; Morey, D. F., 2014.
10. Drake, A. G., 2011; Drake, A. G., M. Coquerelle and G. Colombeau, 2015.
11. Kruska, D., 1988.
12. Holden, C., 2006; Mason, I. L. ed., 1984; Sheldon, J. W., 1992.
13. Storer, T. I. and P. W. Gregory, 1934.

14. Actually, if almost any kind of dog was raised in the way the Russian foxes are, they would likely be much more psychologically disturbed than the foxes. Here is a 2012 description of the facility by Ceiridwen Terrill, who visited the fox farm in person and shared video of the facility on the internet. This is how the foxes are housed, and this is for their entire lives after they are weaned and separated into individual cages: "The fox farm in Novosibirsk, Russia, site of the most famous domestication experiment in the world, currently houses 3,000 silver foxes. Each open-air wooden shed holds 100 or so animals in [individual] adjacent wire cages, each measuring one cubic yard. All wire above, below, and on both sides, the cages have no toys, no stimulation of any kind, except for a cast-aluminum food tray that rotated like a lazy Susan on an axis so that food can be placed on the tray from outside the cage." Each fox had only one cubic yard (a space about 3 feet/one meter square by 3 feet/one meter high) of open mesh wire cage, elevated from the ground with no den or even a board to rest on. The foxes have nothing to do, except when they are tested or fed and no play with other foxes. Testing their reaction to people took a couple minutes per fox during which they are not touched, and was not done every day and not their entire lives. Except for public relations opportunities, they also are handled very little, just the minimum required for testing. The facility is of course clean, and the foxes get good nutrition and health care when needed, but otherwise these conditions are very similar to some of the worst puppy mills. Yet in the pictures and videos, one can see the foxes actively soliciting attention from people, both familiar and unfamiliar to them, wagging tails, pawing at the wire to get closer, and hanging relaxed in people's arms. Dogs that were raised only isolated in cages from birth, especially small pens where they cannot move about freely and get exercise, and that are handled the bare minimum required to care for them, often become very shy of unfamiliar people (which can transform into defensive aggression). When removed from their familiar pen, isolated under-stimulated dogs are typically highly stressed, nervous, and hyper-sensitive to touch and sound. The foxes therefore are not 'dog-like' in their 'friendly' or tamed behavior. They seem almost compulsively solicitous and abnormally resistant to the adverse effects of environmental deprivation.

15. Johannes, J., 2003; Oppenheimer, E. C. and J. R. Oppenheimer, 1975; Pal, S. K., 2008.

16. Anonymous, 2012.

CHAPTER 4. THE DNA STORY OF THE DOG

Modern molecular genetics provides such large quantities of precise data that there is a tendency to overlook its limitations. —R. A. Nichols (1996, p. 365)

Genetics is a complicated science. Genetic analysis is based on mathematical formulas that in most cases can only be solved in a reasonable time by using computers. It takes some study to grasp even the fundamentals. There are so many facets to analyzing DNA for different purposes it would take a full college level course just to explain the methods, why they are used for particular applications, and what sources of error they contain. To understand the formulas and why geneticists use the ones they do, you need to know calculus and statistics. I know basic statistical principles, but not calculus, so I can only address a few major points about some types of information in the formulas used to analyze dog DNA that affect the conclusions. Using as little jargon as possible, I will concentrate on the two most important things for the dog ancestor question: Dating the time of the split between dogs/wolves and their genetic closeness. In the last decade, dog gene sequencing has progressed rapidly thanks to the dog's usefulness as a surrogate for searching for the genes responsible for human diseases, many of which are also present in dogs.[1] This similarity in diseases opened up funding opportunities than would not have been available for research relevant only to dogs. Because purebred dog breeds have small, closely related gene pools started from a small number of founders, there is less variability between individuals within a dog breed than there is between individual humans. This makes it easier to locate target genes with mutations that cause disease. Dog DNA databases now contain partial sequences of several thousand samples and dozens of complete genomes. By contrast, the wolf database probably has several hundred samples and a handful of complete genomes. There are also a large number of coyote samples and smaller numbers of samples for other canids, with some portions of their DNA reported. At this accelerating rate of progress, in another decade or two there may be more answers than questions about canid genetics.

THE STATE OF DOG GENETICS

The only irrefutable evidence that the wolf could be the dog ancestor is their extremely close genetic relatedness.[2] All four of the well-known dog geneticists that I asked "Could this genetic similarity be due to the dog and wolf sharing a recent ancestor?" answered "Yes." The only ones who actually said this in a scientific paper were R. K. Wayne and E. Ostrander. In their 1999 paper they concluded "So, it looks like the dog descended from the wolf or a recent common ancestor." However, they never again mentioned

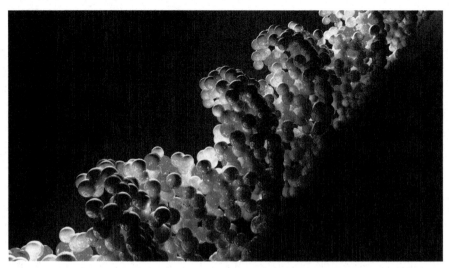

FIGURE 4.1. *A scientifically accurate depiction of a section of a DNA strand showing its structural complexity. Interacting portions working together, called "genes," can be dispersed over the strand making gene detection and analysis extremely complicated.*

a possible alternative to the wolf; nor was this discussed in any other dog genetics paper until 2015. The wolf origin hypothesis is so entrenched that alternatives are not even considered, especially since, as discussed below, the most recent estimates of the time dogs and wolves separated are only 40,000 years ago or much less. Wolves and dogs are more similar in much of their DNA than most subspecies are to each other. How, then, can it be questioned they are not the same species? The answer is that not enough is yet known about the genetic differences between closely related species to make definitive conclusions about species boundaries or to resolve their relationships.[3] Results are especially questionable when only one or a few genes are compared between the species. The popular conception is that genetics, like chemistry or geology, is a hard, well-established science that reaches exact conclusions using unbiased experiments and direct observations of some basic elements of Nature (for DNA, the four base codes). In reality, genetics is a very young science, still in its formative and exploratory stage. It is full of subjective elements.

Perhaps the general feeling of certainty about genetics comes from the undoubted reliability of the DNA "fingerprint" markers used to determine parentage and identify individuals from tiny DNA samples, as popularized in so many crime stories. Beyond the mechanics of sequencing the DNA and using it for identification, much of genetics at this point in its development consists of preliminary research. The conclusions are founded on limited data and analyzed using complex mathematical formulas in computer simulations, not from actual real-world cases. Simulations are necessary because real-life data is not yet available for most things. In many cases actual data is impossible to get because of the extremely long time frames over which evolutionary changes (mutations in DNA) occur.

Outside of a few things like using DNA fingerprinting to determine an individual's identity or parentage, or whether or not a subject belongs to a specific population of a species (possible because members are usually relatives), results announced by geneticists must be taken with variable confidence levels. Results based on fragments of specific genes, or a few genes, have the lowest level of certainty. The highest confidence is of course when whole genomes are compared. Currently very few complete animal genomes have been sequenced and published, and then only one or two specimens per species (two for dogs, a boxer and a poodle; no wolf complete genome published yet, but this is in process).[4] Therefore, most results are based on varied portions of the DNA. Several dog geneticists are currently sequencing the whole genomes of hundreds of additional dogs, including dingoes and aboriginal village dogs, and wolves. Soon we will know much more.

SOURCES OF DOUBT

The general feeling of certainty about genetics is encouraged by the way study results are reported. Often the research conclusions published in science journals are presented without mentioning the possible shortcomings of the data analysis or potential alternative conclusions. Most readers of science journals are aware of the sources of vagueness and the possibility of alternative conclusions (ambiguity). However, popular press news reporters who write about new discoveries for the public usually are not aware of the underlying uncertainties. Unless the geneticists make a point of telling reporters their conclusions are preliminary, their results are presented as if they are highly reliable, as certain as the results of a chemical or spectrographic analysis. As pointed out previously, the biological part of Nature is messy and complicated, making all conclusions tentative to some degree. DNA is incredibly complex. The technology and mathematical tools needed to decipher it add another layer of complication.

The DNA information provided by the human-designed (and therefore fallible) computer programs has to be interpreted by people who choose what data may be important and often have unconscious biases. The programs themselves depend on mostly estimated parameters. There are many variables affecting almost everything in biology, and in genetics some of the variables and their effects are still unidentified. One important variable in determining a species' DNA is the length of sequencing the results are based on. Naturally, the smaller the sections sequenced, the greater the uncertainty of the conclusions. The results from studies based on only portions of DNA tell pieces of the story that are true as far as they go, but conclusions drawn by extrapolating from pieces always have an associated degree of ambiguity. If the study used only 300 base pairs of a DNA strand from one gene to investigate questions such as two species' relationships, the results are much less reliable than those of a study that included 1,000 base pairs or the whole gene. Results based on segments from multiple genes or whole genomes are much more certain.

SPECIES DETERMINATION

Genetic "distance" between species or subspecies is determined by counting the number of differences in the same sections of a gene in different species or separate populations (subspecies) within a species. Mitochondrial DNA (mtDNA) has frequently been used to estimate the genetic relationships between species. Mitochondria are the organelles inside each cell that make the body's energy. Nuclear DNA (nDNA) provides instructions for the organism's body and inherited behavior. There are a lot of mitochondria in each cell and some usually survive even in very old bones, so mtDNA is often still intact in samples that are otherwise degraded through decay or fossilization. Also, mtDNA is simpler to work with because it consists of only one strand of DNA, which replicates without mixing, rather than the two intertwined strands present in nuclear nDNA, which exchange sections with gametes (called recombination). Because mitochondria replicate more often than the nucleus of cells, and mutations (changes in base pairs) mostly happen during replication, they accumulate in mtDNA as much as 10 times faster than nDNA, making differences more obvious.

The main drawback of mtDNA is that it is inherited only from the female line (well, rarely some paternal mtDNA is passed on and mtDNA types can differ between tissues[5]), and so it represents only half of the story about a species. Some of the male side of the species history can be determined from the DNA in the male Y sex chromosome. The genes on Y chromosomes do not recombine during replication with the paired female X chromosome, and so they are a true indicator of direct male gene line inheritance.

Wolves and dogs have only 0.02% different sequences between them in mtDNA. Their next closest relatives, the coyote and golden jackal, differ from the dog/wolf mtDNA by about 7.5%. For comparison, the average mtDNA genetic distance between species in general is at least 2%, so the mitochondria of dogs and wolves are genetically more similar than most subspecies are to their main species population.[6] In addition to the very small mtDNA genetic distance, there is another reason often given for claiming dogs descended directly from wolves. In a diagram of mtDNA haplotypes of dogs and wolves, some wolf types are interspersed with the dog types, and one or two mtDNA types are present in both a few dogs and wolves. Figure 4.2 is a diagram showing this interpolation of haplotypes. The dog mtDNA haplotypes that are different from wolf vary by only one or two changes, so they are very close.

Many authors have cautioned that gene trees (diagrams of related genes such as the mtDNA diagram) are not species trees, and that mtDNA is not a good measure of overall genetic differentiation.[7] One of the main reasons for this caution is that mtDNA often crosses species reproductive barriers when a female mates with a male of another species. If the resulting female offspring bred with a male of her father's species, in a few generations the descendants would be "perfect" genetic specimens of the father's species, except their cell energy organelles were originally in another species. When closely related species pairs (two species that recently diverged from the same ancestor) have ranges in the wild that intersect, some hybridization is expected. In fact, as more species and more specimens of each species are DNA sequenced, formerly unknown instances of hybridization, both ancient and modern, have turned up. Hybridization between species is greater than previously imagined.[7]

Another support offered for the wolf origin hypothesis is that some dog and wolf mtDNA haplotypes are, in addition to being generally similar, interspersed with each other on diagram trees of relatedness, a possible indication they are recently descended from the same maternal line, i.e., belong to the same clade. In a couple cases the dog and wolf actually share a type (they are identical). Figure 4.2 of wolf and dog mtDNA types from Wayne & Ostrander (1999) clearly shows this. As usual, conclusions about dog/wolf mtDNA have been biased by the assumption that the wolf is the direct and recent ancestor of the dog. Any similarities are automatically ascribed to the supposed maternal wolf ancestors. In fact, the dog/wolf mtDNA types are much more separated than those of some other species that have a wild and domesticated population. Next to the dog, one of the oldest domesticated species is the goat, which is thought to have been domesticated about 10,000 years ago. S. Naderi, et al. (2008) reported that all of the mtDNA clades found in domestic goats have also been found in the ancestral wild goats (called bezoars) of the Central Iranian Plateau, the Southern Zagros Mountains, and Eastern Anatolia. In other words, the domestic goat mtDNA types were all interlaced among the wild types from those areas and the domestic goats had no unrelated types. So it seems that the mtDNA comparisons actually indicate the dog

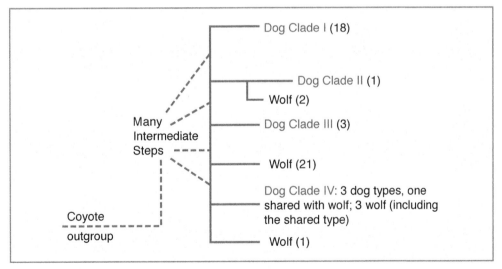

FIGURE 4.2. *Diagram of mtDNA type groups adapted from Wayne & Ostrander 1999 (Fig. 3., p. 250). The coyote was used as an outgroup to start the analysis from a more distant relative. The numbers in parentheses are how many different types they found. This chart illustrates that there are six major groups (clades) of related mtDNA haplotypes in the dogs and gray wolves sequenced. The four main dog clades have been confirmed from later studies with many more dog types found, the majority belonging to Clade I.*

Chapter 4: THE DNA STORY OF THE DOG

and wolf are relatively well separated. It bears repeating that both the dog and wolf could have inherited their original identical or similar mtDNA types from a common ancestor. Also, some later mixing of mtDNA types through hybridization is a virtual certainty. Each gene has its own evolutionary history, which is basically unknowable given current knowledge.

TIME SINCE THE SEPARATION OF SPECIES

Sequence differences between related populations are used to estimate how long ago those two groups (species or subspecies) diverged. The first estimate of genetic distance of how long ago the wolf and dog lines separated, by Vilá, et al. in 1997, was about 135,000 years BP. Since the majority of scientists believe that the dog only separated from the wolf due to association with humans, and modern humans have also been around for about 100,000 years, this seemed much too long ago and was rejected. Later estimates done by several researchers have varied from about 15,000 to 40,000 years BP, a fit for the human time-line. If the date that wolf and dog became separate evolutionary lines were really about 100,000 BP, this would support the hypothesis that the dog evolved naturally.

At present, the dates estimated through genetics are subject to too much doubt to accept any of them without question. However, 15,000 BC is surely too recent if the dog ancestor was the gray wolf, as there are definite dogs existing at 14,000 BC. From gray wolf to dog with so many different traits would certainly have taken more than 1,000 years without strict artificial selection, which was unknown then, before any animals had been domesticated. These separation dates vary so much because estimating the time since two species or evolutionary lines separated is one of the most speculative of all of the techniques in genetics, with margins of error that reach 100%.[8]

There are many sources of uncertainty in DNA analyses. The most tentative conclusions are the dates when two species became separated from a common ancestor and the phylogenetic (evolutionary) relationships among a set of related species. Their uncertainty is due to the assumptions that go into the formulas used to analyze the data.[9] For example, the procedure that estimates the time since two evolutionary lines separated (usually species or subspecies, but sometimes geographically isolated populations) from genetic sequences uses several factors that are themselves estimates. Changing any one of them alters the outcome. The dating formulas may include the following major approximations: The number of mutations accumulated over a time-span, an estimate of the DNA mutation (substitution) rate; an estimate of the time between generations; amount of possible introgression from another species; the estimated population growth rate; and an estimate of how many individuals founded each population. All of these approximations introduce potential sources of inaccuracy. There are other assumptions embedded in the mathematics, and results are also affected by the number of samples used in the estimate. Dating from DNA is at present conditional, not definitive.

THE FOSSIL FACTOR

One necessary variable factor in genetic dating is the DNA mutation rate, which has to be based on the fossil record. Paleontological dating uses unavoidable approximations dependent upon what became fossilized and what fossils have been discovered (which is in turn a function of interest in searching for them and the availability of funding). Paleontologists may also interpret the available fossils differently, an inevitable consequence of Nature's variability combined with individual training and experience. The earliest fossil that matches a living species provides an approximate upper limit in time for that species. The origin date is always provisional because tomorrow an older fossil could be discovered. If the oldest known fossil is clearly already recognizable as a unique species, it must have separated from its ancestor long before the specimen became fossilized. However, since no one can judge how long this prior evolution may have taken, it is ignored. The oldest fossil dates available for two living species provide an estimate of the minimum time they have been unique species. This time divided by the number of different DNA mutations each species has today provides the mutation rate for those species. Mutation rates estimated this way are often used to approximate the time of origin of related species that have no fossil record.

The sub-fossil dog specimens currently available (14,000–26,000 years old) are obviously not "original" dogs. Therefore, geneticists use the approximate mutation rates of the related gray wolf and coyote to time the dog/wolf split. The first few estimated dog/wolf separation dates were based on a coyote/wolf separation of about 1,000,000 BP. Later, this separation date was moved back to 2,000,000 BP, which is closer to the date accepted by a majority of canid paleontologists. P. Savolainen, one of the most prominent dog geneticists, said (2006, p. 133): "[W]ithout more exact paleontological evidence, a precise dating of the origin of the dog will not be possible. . . . To simplify the following argument [about the dating], the 1-million-year date for wolf-coyote divergence will be used, but bearing in mind the possibility that the split occurred up to 2 million YBP

FIGURE 4.3. *A chimpanzee contemplates the skull of a human in this sculpture by Hugo Rheinhold. There are no fossil chimpanzees of an age to anchor the date of the separation of the chimpanzee and human evolutionary lines. The date of DNA estimates has been steadily pushed farther back in time as methods improved and is now considered to be about five to eight million years ago.*

[years before present], in which case the age of the datings should be doubled."

The coyote/wolf split is used to date the dog/wolf split, because it is as certain as anything in paleontology can be that the wolf and coyote shared a common ancestor. This is not the date of the origin of the wolf, but the date that an ancestor common to wolf and coyote (called the most recent common ancestor: MRCA) crossed the Bering Strait and began evolving into the wolf. This common ancestor is believed to be *Canis lepophagus*, which through intermediate species in North America ended up as the coyote. Fossils indicate *lepophagus* lived 4.9–1.8 million years ago. *Canis lepophagus* is also the most likely ancestor of *Canis priscolatrans* (considered one of the earliest small wolves), which migrated from North America into Eurasia across the Bering Strait. The North American *priscolatrans* line evolved, through intermediate species, into the extinct dire wolf, *Canis dirus*, so that line ended about 8,000 BC, when that species went extinct. After entering Eurasia, *priscolatrans* diversified into many species including *Canis etruscus* which could have been the parent species of *Canis chihliensis*, the most likely direct ancestor of *Canis lupus*. So, there were undoubtedly several species along the lines to the modern coyote and the gray wolf.[10]

Since currently there are no fully accepted dogs older than 14,000 BP, and genetic mutations happen irregularly and rarely over an extended time-frame, this would not be long enough to accumulate sufficient mutations among dogs to date their origin. The comparative method is then used. Even severely simplified, the process of calculating the genetic distance between the dog and the wolf is complicated. Geneticists will undoubtedly be displeased by the abbreviated description's insufficiency, but it is intended only to provide an indication of one of the simplest procedures that have been used.

Vilá, et al. (1997), came up with the disputed 135,000 BP dog/wolf divergence date by using the following information: (1) Coyotes and wolves have about 7.5% difference between them in their mtDNA sequences after 1,000,000 estimated years of separation; (2) They divided the 7.5% by the additive 2 million years separation between coyote/wolf (each line evolved for the same length of time so we have to add their times-since-separation); (3) The result is the estimated percentage of changes per evolutionary line per year. Using a calculator this comes out to roughly 0.00000375% of the mtDNA sequences mutating per year in wolves and coyotes. Vilá et al. did not compare the total differences between the dog/wolf mtDNA directly to the number of substitutions between coyote/wolf lines to get a separation time. Instead, they compared the annual rate to the number of mutations found within the largest group of related dog mtDNA sequences, a subset of all dogs. This closely related Clade I (Figure 4.2) is assumed to have started from one maternal ancestor. This method is intended to estimate how long, given the assumption of a single wolf female founder for this clade and 0.00000375% changes per year it would take to develop the number of different mtDNA types within dog Clade I today. The divergence within this clade was about 1%. By their calculations, this results in the estimate Clade I is about 135,000 years old. They report they tested the substitution rates of dog and wolf mtDNA sequences and they were similar.[11] Some recent papers use 2 million years for the coyote/wolf split date. With the method used in Vilá et al. that would make the dogs look even older.

Interestingly, according to Vilá et al. (1997) the sequence diversity among dog mtDNA haplotypes is almost as high as that of the wolf: within dogs it is about 2.06% average (4.67% maximum), and within wolves selected from different geographic areas about 2.10% average (3.95% maximum). It seems that if the dog originated from the wolf only 20,000–40,000 BP, and their rates of substitution are the same, then to get this much divergence in that time, the dogs either had to be founded by a very large population of wolves with a lot of diversity among them to start with, or the dog has had about the same amount of time as the wolf to accumulate mutations among themselves. But apparently this "common sense" conclusion is not valid, or surely the geneticists would have mentioned this possibility. None have.

THE VARIABLES

P. Savolainen, et al. (2002) estimated the mutation rate for dog mtDNA was approximately one mutation per 24,000 years within a lineage. The substitution of one base pair of DNA for another "letter" of the DNA alphabet is stochastic, not perfectly regular, so the rates given are an average estimate over a stated period of time. This means the change could have happened in the parent of the individual sampled, or 5,000 generations in the past. It was formerly thought that mutations happened like clockwork over time affected only by the DNA replication process. This was called the Molecular Clock hypothesis. The "clock" runs faster in rodents and much slower in humans. For various reasons, mutations happen more often in a given time limit in rats than in humans. But as knowledge expanded, the geneticists discovered many exceptions to this clock, although it is still used for rough approximations. In addition, individual genes can also have different rates of mutation.[12]

Recent studies reveal that even in some related species the mutation rate can be quite different. There is no way to really know at this time if dog DNA evolves at the same rate as wolf DNA, although this is a reasonable expectation given their similarities. For the time being, wolves are the best available comparison for dogs. Applying the mutation rate of one species to another will never result in a perfect match. The species being compared both have accumulated mutations since their common ancestor, through possibly one or more species in each direct ancestral history. Those ancestral species could have had a different genetic evolutionary rate for several ecological reasons (such as the effect of different generation times). Without knowing the ancestor's gene sequences, it is actually impossible to say how many of the total changes happened in each daughter species' history.

G. Larson and D. G. Bradley (2014) discussed the differences in the times reported since dog and wolf separated in articles by A. H. Freedman, et al. (2014) and G.-D.

Wang, et al. (2013). Freedman et al. concluded the dog diverged from the wolf about 11,000–16,000 BP, and that domestication resulted in a 16-fold reduction in early dog population size. G.-D. Wang, et al. concluded the divergence was about 32,000 BP, and the population bottleneck resulting from domestication was not significant. Why these major differences? As Freedman et al. (2014) point out, little is known about the dog-specific mutation rate, and those used by the various researchers differ by an order of magnitude (10 times). Larson and Bradley (2014, p. 1687) say "The use of the entire range of rates therefore results in a credible interval of the origin of dogs from 9,000 – 34,000 years ago, certainly in greater agreement with the archaeological estimates [than the 135,000 years in Vilà, et al. (1997)], but still lacking precision."

The estimated number of original founders is another important variable and source of potential error in dating formulas. Founder number works this way: As the number of founders goes up, the estimated time of separation goes down. If an evolutionary line started from many original sets of DNA that already had some differences, it would have taken less time for the genetic variations present today to accumulate. In the case of the dog, the estimates (almost guesses really) of how many mtDNA (female) founders there were for the population that became domesticated, has varied from four to over 40. The original 135,000 BP estimate was based on the assumption of a single founding female for that clade of related mtDNA types, and no subsequent hybridization with wolves. As dog DNA studies progressed, it became apparent from the mtDNA diversity present in dogs that there were probably many founding females, perhaps as many as several dozen, and that there had also been some rare hybridizing with wolves. The dog mtDNA variations found so far all fall into just four clades, so there were at least that many founding females. There could have been many more that all had similar mtDNA. More recent studies have postulated 20 – 40 founding females. J.-F. Pang, et al. (2009) looked at the mtDNA control region of 1,543 dogs. They concluded the average sequence distance to wolf ancestral haplotypes indicated dogs originated 5,400 –16,300 years ago, and that there were at least 51 female wolf founders.

It is highly likely that female proto-domestic dogs, wolf or dog, adapted to living in the human-centered environment, had their pups near the humans. Most pups would grow up accustomed to scavenging around people. Like the tundra and forest wolves, these proto-dogs were more likely to breed within their ecological group, thus limiting the number of female founders.[13]

However, proto-dogs probably sometimes formed a bonded pair with wild conspecifics during breeding season, and the wild ones adapted to being around humans because they would not abandon their mates. This would expand the number of genetic founders overall, but would only add to the mtDNA diversity if the wild mate was female. The problem with estimating from current genetic diversity how many founders there were for dogs, is that the actual magnitude of differences present among modern dogs is indeterminate. The data we have are necessarily based on a relatively tiny sample size from the 700 million dogs alive today.[14]

Genetic introgression from the supposed ancestral species breeding to the daughter species is another variable that can strongly affect the estimate of how long ago the two separated. The common default in dating from mtDNA gene sequences has been to assume complete genetic isolation of the population. This is rarely the case as most closely related species that live in contact with each other sometimes make 'mistakes' choosing mates and genes cross the species reproductive barrier. After domestication, as dogs spread throughout the world, dogs and wolves have on rare occasions interbred. This should be taken into consideration because it changes the conclusions.

A recent dog/wolf separation estimate using mtDNA cytochrome b, a gene commonly used for species identification, illustrates how including introgression from another population or species can affect the results.[15] In their 2011 paper P. Skoglund, et al. estimated the dog/wolf divergence time two ways. First, assuming complete isolation without gene flow, they concluded the dog population started about 3,500 generations before the present, corresponding to approximately 10,000 years ago (95% confidence interval: 9,000 – 13,000 years). A 95% confidence interval [CI] means there is a 95% chance the true result would fall somewhere in that range. Then they included a low rate of gene flow from wolf to dog in the formula (but of course genes could have gone either way, wolf to dog or dog to wolf, further confounding analysis of results). This increased the estimated time since divergence to about 30,000 BP (95% CI: 15,000–90,000). The computer programs used to run DNA analyses repeat the process for each formula several hundred to a few thousand times. The outcomes are not identical because the program uses several different values for the variables. Results are reported as the most frequently obtained outcome (the mathematical mode), or the median in a range of results. P. Skoglund et al. (2011) gave the mode of 30,000 BP for the result of their second formula, and not the median of the CI, 37,500 years BP. The DNA reports nearly always include modifiers such as "likely" and "approximately" because of the variability inherent in this type of analysis. The margin of error for genetic dating is often 100%, which would be unacceptable in most other types of analyses. When the CI is 15,000 – 90,000 years BP, a six-fold difference, that is an indication some techniques of genetics are still in a developmental stage.[16]

THE REST OF THE STORY

Nuclear DNA (nDNA) is incredibly complex. Over millions of years, billions of changes occurred requiring even more changes and adjustments in other parts of the genome. How genomes work, how they interact with the environment to produce organisms, is still largely a mystery. The nDNA consists of protein coding genes, genes that function to direct development, "start" and "stop" codes for replication, transcription factors that indicate what parts of the gene are replicated, plus lots of other things including segments for which the function has not yet been figured out. The mtDNA mutation rate is 10 times higher than nDNA, which is why it is used for so many types of species relationship estimates. Nuclear genes are much more highly conserved (slow to change) than mtDNA because the coding parts — the actual protein genes — are templates used to build animal bodies. Some slight variations in the code make no difference. But, sometimes the change of just one letter of a base pair alters the resulting protein enough so it is not functional, thus replication has to have very few mistakes (mutations) to keep the organism viable.

In structural nDNA genes (those that code for proteins or RNA, the messenger particles), the mutation rate is very low, on the order of about 10-5 (or 0.00001) mutations per gene, per generation. This is why approximately 94% of the dog protein coding DNA is the same as that of mice, rats and humans, and every human on the planet shares 99.99% of these common gene sequences. However, the actual protein coding genes comprise only about 1.2% of the human genome! Most nDNA consists of what used to be called "junk DNA" because it did not code for a structural protein. According to The Chimpanzee Sequencing and Analysis Consortium (2005), the differences between humans and chimpanzees in "indels," sections of DNA that have been inserted or deleted from genomes, account for about a 3% additional difference. Copy number variants, where species have a different number of copies of the same genes, is another 6.4% difference. The number of copies can affect how the gene is expressed, making more or less of a product. Adding all these bits up indicates there is at least a 10.4% difference between human and chimp genomes, not counting the non-coding regulatory elements. We know that the non-coding regions contain developmental and regulatory sequences that tell the coding genes where and when to start and stop working. Like the protein genes, even a small difference in the timing of these activities can make a large difference in the end product, the individual animal, so these regions are also fairly highly conserved. Small changes in these powerful regulatory genes, or in the development genes turned on by them, represent the major source of evolutionary change.[17]

Many of the big questions of how DNA works remain to be answered. These are questions about how the sequences within one gene interact with each other, with other genes, with other cellular elements, and with the environment, to result in an adult organism or even in the development of an inherited disease. When people are discussing how the wolf turned into the dog, frequently the dismissive statement is made that all it took was "just" some minor changes in development, but this is exactly the main difference between us and a mouse! As pointed out previously, dogs are not, as so many have concluded, paedomorphic wolves with development arrested before the adult stage. The dog is merely different, and the differences could be due to natural adaptive evolution. How the genes translate into organisms is one of the most complex and therefore least understood subjects in biology, but progress is being made. Evo Devo ("Evolution-Development") is the study of how the process of development into an adult organism changes over evolutionary time to create novel species. This is a fairly new subject, dedicated to unraveling the mysteries of how genes are regulated.

Most somewhat related organisms, such as all mammals or all birds, are extremely similar as embryos and go through many of the same early forms as they develop. For example, all mammal embryos at some stages in their development have gills or gill pouches and similar overall shapes. Evolution, a change of shape or behavior in the adult organism that starts a new evolutionary line that could become a species, is not caused by changes in coding genes, but by changes in the expression of the coding genes. The main engines of evolution are the differences due to mutation/substitution effects on the expression of the genes.

Homeobox genes are the genes that guide development. They indicate "what" is constructed and "where." Homeobox genes are regulated by DNA timing sequences that govern when, for example, the gene for appendages is turned off and on. One gene can be regulated by many separate switches and that gene, or portions of it, may be used many times for different things. A switch can be several hundred base pairs of DNA long, and facilitate the production of a half dozen to twenty or more different proteins produced by the same gene. The genes that code for legs in a dog also work to create legs and arms in a human.

Scientists have manipulated the development genes in fruit flies, and caused legs to grow out of eyes, or two abdomens to be formed. Glitches in the what, when, and where of gene function are responsible for two-headed calves and snakes with legs. Developmental genes across species are essentially identical but can be expressed in a nearly unlimited number of ways. This is one reason the genomes of so many very different looking species can be so similar.[18]

One last potential source of error in any genetic test that relies on samples from a single type of tissue, for instance only from skin, or any individual body part, is genetic chimerism. This condition, where the genome of some parts of the body or some cells are different from others, results from merging of two fertilized eggs at the start of embryo development or from mutations in certain cell lines. As more instances of genetic chimerism are

FIGURE 4.4. *The common chimpanzee and human genomes differ by only about 10.4%. This reality makes clear why small changes in developmental regulation and other non-coding regions result in very different organisms, physically and behaviorally.*

Chapter 4: THE DNA STORY OF THE DOG

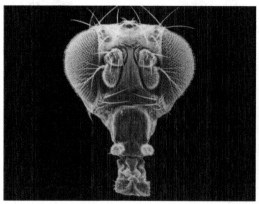 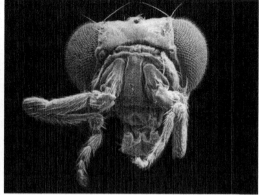

FIGURE 4.5. *On the left is a normal fruit fly head, and on the right one with legs growing where antennae would normally grow after the homeobox gene was manipulated.*

revealed during genome sequencing and testing for organ transplant compatibility, questions arise about the reliability of single samples for representing the organism's main genome. In one study using various samples from the autopsies of 75 women, they found 13 of the female kidneys, 10 livers and 4 hearts had some male cells. There was no apparent failure of these organs, which had functioned normally.[19]

SUMMARY

The above examples of how much the conclusions from genetic analysis depend on various estimated factors (actual values unknown) should serve to demonstrate that at this stage of the development of genetics as a science, the results of most studies should be considered preliminary. With the exception of DNA fingerprinting for identifying an individual and their close relatives, and a few well-accepted DNA associations with particular diseases, genetics is not yet a completely "objective" science just because it is based on mathematics and uses complex machines and computers. The technology is impressive, but using computer programs that require the input of human-estimated values based on necessarily incomplete information cannot result in anything close to a certain or objectively determined result. In addition to the variability inherent in results based on the more or less reliable "guesstimates" included in the computer programs, in the case of dogs there often is a biased interpretation of results based on the belief that the dog is a domesticated gray wolf. Every study, no matter how small, is a welcome addition to our accumulated DNA knowledge base about the mutation process, how DNA differs after population splitting (how species evolve), and how species can be defined by genetics.

However, because DNA study results currently are based on a relatively small proportion of any species' population, the results cannot be considered truly typical for any entire species. When reading the newest revelation about dogs from a genetic study, one must mentally insert the terms "for the samples sequenced" and ask one's self what estimates in the analysis could have affected the conclusions. The one undeniable basic fact from the totality of dog/wolf DNA research is that dogs and wolves can be separated using only genetics. Several types of DNA can be used in analyses to clearly separate dogs, wolves, and their hybrids without the researchers being aware beforehand which samples are which species (called a "blinded study" and believed to eliminate any unconscious biases of the researchers interpreting the results). B. M. von Holdt and colleagues (2011) used over 48,000 single nucleotide polymorphisms (a DNA sequence variation occurring when a single nucleotide — A, T, C, or G — in the genome changes) to compare the wolf and dog. They succinctly reported (2011, p. 1296) that "Our result shows dogs and gray wolves are genetically distinct." Thus, while on the surface the dog/wolf genetic similarity could be construed to uphold the wolf as the dog ancestor, the truth is that all that can be said with certainty is the dog is the closest relative of the wolf. The similarity could be because the dog and wolf share a recent common ancestor. The main reason the wolf has been declared the direct ancestor of the dog is due to the extreme bias that the dog is not a natural species, but an artificial subspecies created by human activities, direct or indirect, and so its nearest living relative "must be" its ancestor.

My position on the current state of genetics was best stated by E. Birney, head of the consortium of 400 geneticists working on the Encyclopedia of DNA Elements project (ENCODE), which is unraveling the mysteries of the function of the non-coding genome. He recently said: "I get this strong feeling that previously I was ignorant of my own ignorance, and now I understand my ignorance. It is slightly depressing as you realize how ignorant you are. But this is progress." (Hall, S. S., 2012, p. 82). New discoveries are being made every day. The automation of sequencing is rapidly improving and getting faster (thus cheaper). Data bases of sequences, markers, and known genes grow by leaps and bounds, and understanding of the basics of how genes work increases daily. It may take another few decades before genetics has more answers than questions, but the bottom line is that dogs and wolves are not genetically identical and the amount of difference in their DNA, just like the number of differences in their bodies, could indicate that they have been separate evolutionary lines for much longer than the 14,000–40,000 years most frequently estimated from DNA dating.

Hopefully, soon the stories about dogs that DNA has to tell will be fully revealed. I agree with Larson and Bradley (2014) that the on-going project sequencing multiple, complete genomes of modern dogs and wolves is a significant step toward the hunt for dog origins. The mystery will only be fully unraveled by using the newest DNA methods on ancient canid specimens, merging genetics and archaeology. They stated (Larson and Bradley, p. 1688): "By combining the expertise of both disciplines, not only might the extinct population of ancestral wolves be identified, but we will gain an enormous insight into the timing, location, and admixture patterns of dogs and wolves, thus revealing the complex origins of our first and best friend."

NOTES

1. Griffiths, P. E. and K. Stotz, 2006; Ostrander, E. A., 2012; Sutter, N. B., and E. A. Ostrander, 2004.

   What is a gene? As knowledge of genome function grows, the definition of "gene" is evolving. Genes are no longer the discrete entities, packets of linked nucleotides with clear boundaries, that they were originally thought to be. H. Pearson (2006) defined a gene as an identifiable location on the genome that corresponds to a unit of inheritance whose expression is directed by associated functional regions of the genome (the parts formerly called "junk DNA"). The associated regions could be some distance from the coding sequences, even on another chromosome. J. Hey (2001, p. 38) said: "To date, gene is probably our single most common biological term. It is a bit ironic that as our knowledge of genetic material has grown, GENE has become more difficult to define."

2. Brower, A.V.Z., R. DeSalle, and A. Vogler, 1996; Hey, J., 2001; Hoelzer G. A., 1997. Also see L. L. Knowles and B. C. Carstens (2007) for additional discussion of gene trees vs. species trees. They address what they term the inherent biases in species detection arising from when and how speciation occurred, and failure to take into account the high stochastic variance of genetic processes.

   What does "closely related" mean? In evolutionary terms, closely related usually means species that have diverged from a common ancestor recently, from 100,000 to more than a 500,000 years ago depending on the type of animal. When separation of evolutionary lines from a common ancestor was recent for a group of species, often there is what is called "incomplete lineage sorting" of their genes. They have not been separated long enough for each to develop their own unique combinations and some shared ancestral haplotypes are retained. The last common human and chimpanzee ancestor (both the common chimpanzee and the bonobo) was between 4 to 7 million BP, and there is about 1.5% difference between humans and chimps in protein coding genes. According to Polavarapu et al. (2011) the non-coding or regulatory sections of DNA are where humans and chimps differ the most, by about 4%. This is why it is worth repeating that trees of the similarity of coding and mtDNA genes are not necessarily species trees.

3. Cronin, M. A., et al., 1991; Cronin, M. A., 1993; Dávalos, L. M. and A. L. Russell, 2014; Knowles L. L. and B. C. Carstens, 2007; Smukowski, C. S. and M. A. F. Noor, 2011.

4. So far at least one individual from the following mammals have had their complete genomes published and others are in process: Human, mouse, dog, cow, rat, fin whale, blue whale, harbor seal, grey seal, horse, American opossum, hedgehog, gorilla, common and pygmy chimpanzee, Bornean orangutan, Sumatran orangutan, Indian rhinoceros, donkey, platypus, guinea pig, cat, white rhinoceros, wallaroo, and armadillo.

5. Akst, J., 2010; He, Y., et al., 2010; Zhao, X., et al., 2004.

6. This clear separation between species is also not present in some other recently evolved species that radiated from a common ancestor. The Lake Victoria cichlid fish species studied by I. E. Samonte et al. (2007) could not be separated by their genes alone, even though they are considered "good" species with strong adaptations to different niches. The fish were about 0.06% different in mtDNA and 0.015% in nDNA. The authors concluded that this amount of separation equated to about 14,000 years evolution, which is close to the time Lake Victoria re-filled after completely drying out, indicating speciation from a common ancestor happened after the lake filled again. They said (p. 2078):

   The study spotlights the fundamental issue in the investigation of early speciation phases—the problem of distinguishing gene flow from shared ancestral polymorphism. The approach we have chosen does allow us to make this distinction and to come to the conclusion that the 4 tested species have been exchanging genes from their inception at a rate sufficient to homogenize the species genetically. The homogenization affects large parts of the species' genomes, presumably with the exception of the few genes controlling the phenotypic interspecies differences. . . . Because of their compounded and genetically promiscuous history, their phylogeny will not be resolvable, no matter how many genes future biologists will use in their attempts at phylogenetic reconstruction of this adaptive radiation. Similar situation is, in fact, encountered today in attempts at reconstruction of phylogenies of species that have survived past adaptive radiations.

   The Darwin's finches of the Galapagos Islands are also still undergoing adaptive radiation even though the common ancestor arrived there over 2 million years ago. Sato et al. (2011) concluded that there were no specific gene sequences in the samples they studied that could be considered species, genus, or even Darwin finch group specific. As in the cichlids, physical differentiation into species is running ahead of differentiation in genetic systems such as the mitochondrial DNA. The dog and wolf are, I maintain, two species from just such an adaptive radiation from a single, still unidentified ancestral species. Dogs and wolves can be identified by genes alone, and so it seems they must have been separated for much longer than the 14,000 years for these fish species and the youngest species of Darwin's finches. Dogs and wolves have had only rare hybridization events once they evolved into dog and wolf, probably because they were for a long time geographically isolated from each other. By the time they met up again, the likelihood of cross-breeding was much less than in the fish and birds living in the same place.

7. Arnold, M. L., 1997; Grant, P. R., R. Grant and K. Pentren, 2005; Hoelzer, G. A. 1997.

Von Holdt et al. (2011) estimated the amount of dog and North American wolf genes present in various coyote populations, and did not report any dog genes in any wolf sample. The Northeastern coyote population has on average 9.1% (range: 5.2%–12.8%) dog ancestry and the Midwestern/Southern population has 4.4% (range: 1.9%–8.1%). The highest level of dog ancestry in coyotes was identified in Ohio and Virginia, 16.7% and 16.9%, respectively. Another example: Randi and Lucchini (2002) found that all the Italian wolves and dogs in their study could be assigned to two different inter-related clades, and the mixed gene composition of the dog/wolf hybrids showed up because they were assigned to more than one of the clades.

8. Bandelt, H-J. 2007; Bromham L., 2011; Conroy, C. J. and M. vanTuinen, 2003; Cox, M. P., 2008.

9. Dávalos, L. M. and A. L. Russell 2014; Larson, G., et al., 2012; Scally, A. and R. Durbin, 2012; Schenekar, T. and S. Weiss, 2011; Skoglund, P., A. Götherström and M. Jakobsson, 2011; Smukowski, C. S. and M. A. F. Noor, 2011.

10. Tedford, R. H., B. E. Taylor and X. Wang, 1995; Wang, X. and R. H. Tedford, 2008.

11. A recent paper by A. Scally & R. Durbin (2012) came to the surprising conclusion that the human mutation rate is about half what it was assumed to be from the calibration derived from the fossil record of chimpanzees and humans. They were able to do this using generations of human relatives in Iceland to get an actual, direct number of substitutions between the relatives. They say that this has 'implications for our understanding of demographic events in human evolution and other aspects of population genetics.' A plausible explanation, given that the phylogenetic estimates for humans and chimps is averaged over lineages extending back millions of years, is that there has been a decrease in mutation rate within the great apes since their divergence from other primates, and the human mtDNA mutation rate slowed even more.

12. Akst, J., 2010; Bromham L., 2011; Laidlaw, J. et al.,, 2007; Nabholz, B., S. Glémin, S. and N. Galtier, 2009.

    For comparison to the mtDNA rate, the rate estimate of 3.8×10^{-9} substitutions per site per year for canid ribosomal DNA was provided in Aggarwal et al. (2007). That comes out to about 0.0000000038 changes per individual DNA base per year for that gene.

13. Assortive mating or genetic separation of populations due to ecological (niche) differences is discussed further in Chapters 5 and 6.

14. Gompper, M. E., 2013; Hughes, J. and D. W. Macdonald, 2013; Ritchie et al., 2013.

15. E. K. Rueness et al. (2011) reported the following results: The Cyt b gene

of the mitochondria is often used to distinguish among mammalian species, and a greater than 5% divergence is typically observed between recognized species. The Himalayan wolf diverges from the Holarctic gray wolf by 1.2% and the Indian wolf diverges from the Holarctic grey wolf by 2.5% (using 332 base pairs of the D-loop of the mitochondria). The divergence between C. aureus lupaster (formerly considered a subspecies of golden jackal, but now considered a subspecies gray wolf, C. lupus lupaster) and C. lupus is 4.0% for the equivalent sequence fragment, while it is 2.4% between C. a. lupaster and the Himalayan wolf. Wayne (1997) estimated an mtDNA molecular clock applicable across Canidae as 1.3 – 1.7% changes per million years. Rueness et al. stated that estimates of evolutionary time based on restricted data tend to be highly uncertain, but based on sequence divergence they found it is reasonable to consider C. a. lupaster as a distinct taxon within the grey wolf species complex.

16. As discussed in M. P. Cox (2008), here is another conclusion about one variable (source of error) in estimating time since the most recent common ancestor (TMRCA) of two genetic populations. N0 is the number in the equation used to calculate the TMRCA that represents the number (N) of individuals in the population at the time of the split, time 0. Cox says (p. 347):

As noted, molecular dates calculated for a constant-size population with N0 = 10^3 have a 34% error rate when using confidence intervals that assume a normal approximation (a bell curve distribution) and a 24% error rate when using quantile based confidence intervals . . . Here, **the error rate is seen to increase with the effective size of the population**. For N0 = 10^4 (i.e. the global effective size of modern humans . . .), confidence intervals generated with the normal approximation have an error rate approaching 75%, and confidence intervals inferred directly from the quantiles of l have an error rate of 44%. **As a general rule, confidence intervals do not contain the true TMRCA for a substantial proportion of data sets, and this error increases with the effective size of the population under study.** (Emphasis added.)

Of course, the actual number of individuals in the populations that started new evolutionary lines is not knowable. They are calculated using another formula that is based on the estimated time of the split and the types and number of mutations projected to have taken place since the split. Note there are two generations of "estimates" in the formula: The resulting number is an estimate based on two other estimates. Since populations rarely stay at a constant size and can fluctuate widely, assuming a constant population size since the species split is another major source of potential error. It is no wonder that the confidence intervals (the margin of error: upper and lower limits of what the actual number could be) are so huge! The margin of error of course does not mean any result was compared directly to a known example of a "real" or actual date of separation because there are none. The error rate is an estimate of the highest and lowest "real number" based on varying factors in the formula.

17. Arendt, M., et al., 2014; Cohen, J., 2007; Gingeras, T. R., 2007; Griffiths, P. E. and K. Stotz, 2006; Hall, S. S. 2012; Ptashne, M., 2013; Willingham, A. T. and T. R. Gingeras, 2006.
18. Duboule, D., 1995; Gehring, W. J., 1998; Ruddle, F. H., et al., 1994.
19. Giorgi, E., 2015; Koopmans, M., et al., 2005.

CHAPTER 5. NATURAL DOG BEHAVIOR?

To some extent . . . comparing dogs to their wild ancestors can be illuminating—but when the wolf is taken as the only available point of reference, our understanding of dogs suffers. . . . By analyzing the dog as its own animal rather than as a lesser wolf, we have the opportunity to understand it—and refine our dealings with it—as never before.—J. Bradshaw, Dog Sense: How the New Science of Dog Behavior Can Make You a Better Friend to Your Pet. (Basic Books, 2011, p. 2)

The Chapter title has the term "natural" with a question mark because we know almost nothing about natural dog behavior, especially intraspecific (dog-dog) social behavior. In this context natural means behavior that has not been directly influenced by humans. Because dogs are generalist predator-scavengers adapted to roaming over large areas, even when groups are kept in large enclosures, there is an effect on social behavior. Furthermore, what little we do know has, until very recently, been misinterpreted.

Wolf behavior has been almost universally considered the ancestral template for dog behavior. Therefore, dog behavior is usually analyzed to see where it differs from wolf behavior, and any difference is attributed to domestication-caused changes. Like the skull differences discussed in Chapter 2, these are also labeled "domestication effects." For example, J. Hecht and A. Horowitz (2015, p. 6) say ". . . [D]ogs' artificial selection history is explanatory of important differences in the behavior of wolves and dogs." Some behavior of today's mixed and purebred dogs has of course been altered by artificial selection for human-desired traits, but until we can figure out what the natural, ancestral template is, we cannot know what is merely different from the wolf and what has changed due to domestication effects.

While dog behavior *is* very similar to wolf behavior, it is just as similar to coyote behavior, jackal behavior, and even red fox behavior. In fact, most behaviors are present in all of the Canidae species. For instance, they all scratch themselves with their hind feet, growl, chase fast moving things, and pin small objects and prey down with their front feet. Very few individual nonsocial behaviors are specific to one or a few species, although the behaviors may be expressed a bit differently in each species. For example, all wild canids use elimination to "mark" their territories, but some use urine, some feces, some mark in open areas, and some only near rocks or bushes. Other behaviors differ only in how often they occur, being rare in some species, common in others. This Chapter will briefly address the few certain behavioral differences between the wolf and dog, some short-comings of published behavior research, and outline the behavior of free-ranging dogs, which reveals the one misidentified major difference between the social behaviors of the two species: The dog is not adapted to living in packs.[1]

DIFFERENCES BETWEEN DOG AND WOLF BEHAVIOR

One of the basic differences between the dog and the captive, tamed wolf seems to be in "bidability" or the willingness to comply with requests for specific behaviors from a human. Most domestic dogs excel in bidability compared to wolves, which offer more resistance to taking direction. Another behavior that appears to generically differ between dogs and wolves is that wolves are much more difficult to train to inhibit their behavior, to comply with commands like "stay." These differences are not universal, as there are races of dogs, such as the African Basenji and some terriers, that also are not in general biddable and have difficulty with inhibition.[2] When looked at closely, most differences between dogs and wolves become a mosaic of variability and similarity, from which only a few general tendencies like bidability can be termed "typical" for each species.

There are also a few differences in the development of behaviors between domestic dogs and wolves. The primary (sensitive) socialization period is the stage from when canid pups can see and hear to when their ability to form attachments to other beings starts to fade. This period was defined for dogs by numerous experiments over many years that isolated puppies from intimate human contact for variable numbers of days or weeks, then tested them for how completely and quickly they adapted to human handling. In dogs this period generally starts at 3 weeks and ends at about 12 to 14 weeks. In wolves the sensitive socialization period starts at about 14 days and ends at about 6 weeks.[3, 25] After the sensitive period, socialization to humans and conspecifics is still possible, especially for domestic dogs, but the older the canid gets, the more extended effort it takes. However, during this period, amazingly little positive interaction with humans is necessary for a domestic dog pup to accept humans and to trust them. J. P. Scott and J. L. Fuller (1965) found that some (they did not test all) puppies almost completely isolated from human contact can be successfully socialized after 8 weeks of age with as little human contact as two 20-minute periods per week. Wolf pups require much more intense socialization starting by about 2 weeks of age (just after the eyes open) to become comfortable with human handling. Socialization to people past the sensitive period rarely results in the same level of trusting relationship possible with early socialization, and often does not transfer to unfamiliar people, especially in wolves.[4]

Cognition (conscious mental activities such as thinking, understanding, learning, and remembering) research on dogs and wolves has recently expanded. The "point-gaze cue" studies where dogs or wolves are tested to see how well they follow a human's gaze or point to a hidden treat have been done by several different investigators.[5] The general consensus is that dogs, even young puppies, are much better at this task than wolves. The Russian tame foxes were also very good at following point cues, similar to the dog's ability. Of course, the marked difference between dog and wolf results is usually attributed to domestication effects, to selection in dogs for attending to and cooperating with humans. However, in a follow-up to an original pointing study done with juvenile wolves and dogs, in which the wolves did not perform nearly as well as the dogs, M. Gásci et al. (2009) tested the same wolves as adults and discovered they then performed as well as the dogs. Their conclusion was the wolves needed a longer development period or more exposure to people to reach the same proficiency as dogs. This makes sense, as wolves are big, about twice the body size of natural dogs (aboriginal village dogs and dingoes), and require a longer period of physical and mental development before they can be independent. The bottom line is both dogs and wolves can use "referential" (indicating something) signals from both conspecifics (head turning) and humans (gaze, head turning and pointing). The dog's seemingly greater aptitude in following human gestures may be due to an increased ability to generalize from conspecifics to humans, and not a "new" behavior that emerged due to domestication.

THE QUESTION OF DOMESTICATION EFFECTS

Almost all studies directly comparing captive, tamed wolf with dog behavior have one common drawback: The differences found may represent domestication effects, but since they all used modern pure breed or mixed breed dogs they tell us nothing about possible original wolf/dog differences. They do tell us how some current wolves and dogs react in the test situations, which also may or may not actually be testing for the behaviors they are supposed to be testing for, but critiquing them is beyond the scope of this Chapter. The best studies start by raising wolves and dogs in as similar circumstances as possible to reduce environmental effects. Some of the earliest such experiments in wolf/dog comparison were by H. and M. G. Frank (1985). The dogs the Franks used were Malamutes, a working dog selected specifically for inhibition of instinctive behavior and obeying commands. They raised wolf and Malamute pups in similar ways and then tested them at various tasks and found large differences in certain behaviors, especially in self-control (much higher in the dogs) and problem solving (much higher in the wolves).

The malamutes were a good choice for the comparison if the goal had been to determine the result of artificial selection on dog behavior, but not a good choice for revealing natural dog behavior. Using modern pure breed or mixed breed dogs makes the comparison suspect, at least with most types of dogs (working dogs such as the Malamute, some hounds, bird dogs, retrievers, companion dogs), because their behavior has been modified by fairly recent strong artificial selection. If Basenjis or terriers were raised in the same way as wolf subjects, a comparison of their behavior would probably have different results for self-control and problem solving. In fact, that is just what the first long-term study of dog behavior by J. P. Scott and J. L. Fuller (1965) found: Terriers and Basenjis were much harder to condition to "stay" than the beagles, spaniels and Shetland sheepdogs, but much better at solving problems.[6]

S. Marshall-Pescini and J. Kaminski (2014) provide an excellent synopsis of the various hypotheses of wolf domestication and of the state of comparative canid cognition studies. They call for choosing subject populations more carefully, so the past experiences of the dogs and wolves are equivalent, and they point out the need for more standardized methodologies. Many of the current studies are not directly comparable because the methods were different in ways that could affect the results. Because today's wolves are just as far removed from the wolf-dog ancestor as the dogs, they point out "... a certain amount of caution appears necessary when concluding that domestication is directly responsible for all (author emphasis) the potential differences observed between these two species." *(p. 21)* They go on to point out the need to study more wolf populations because, given the high intraspecific (within the species) variability of canid behavior due to ecological factors, there may be greater social flexibility among wolves than currently realized.

Another canid/human social interaction study by A. Miklósi, et al. (2003) indicated dogs, but not wolves, would look to a human (for assistance or further information?) when presented with an impossible task. They subjected 7 wolves and 9 dogs to a test that required pulling a rope that passed through a wire barrier to get the meat treat tied on the other end. First, each subject was given 6 training trials pulling the rope, and all learned to do this with their feet or mouth. Then the rope was anchored on the opposite end so it was impossible to pull it. While the handler stood still nearby, the subjects were observed for 2 minutes during the impossible task and every time they looked at the handler was recorded. Of the 9 dogs, 7 looked back at the owner, in contrast to only 2 of the 7 wolves. The dogs looked back sooner in the trial (average one minute into the trial) and gazed at the handler longer than the wolves, which basically merely glanced at the handler. The wolves, in other words, kept trying to solve the problem themselves. A. Miklósi, et al. (2003) combined the results of this rope pull test, with two others also designed to test the ability of wolves and dogs to seek information from humans, into the following conclusion (p. 763):

> Based on these observations, we suggest that the key difference between dog and wolf behavior is the dogs' ability to look at the human's face. Since looking behavior has an important function in initializing and maintaining communicative interaction in human communication systems, we suppose that by positive feedback processes (both evolutionary and ontogenetically) the readiness of dogs to look at the human face has lead [sic] to complex forms of dog-human communication that cannot be achieved in wolves even after extended socialization.

In a later experiment, I replicated this rope pull test with 6 well-socialized New Guinea dingoes that I had raised from 8 weeks of age or birth. Despite letting the dingoes continue to try to pull the tied rope for much longer than the 2 minutes of official time, not one of them ever looked up at me. This was even less human-directed gazing than the wolves demonstrated. These dingoes, commonly called New Guinea singing dogs, are noted for their penchant to gaze directly into the eyes of people for extended periods, several seconds at a time. It unnerves some people, who believe such a direct gaze indicates the dog is signaling "dominance" and potential aggression, which is not the case at least with this subspecies. Those intimate with these dingoes call this the "Singer Look." Expressed as a personal experience, an elderly lady meeting a Singer's gaze for the first time commented to me "It's as if they are looking into your soul." It seems these dingoes are curious when meeting new humans and study their faces, for what clues we can only imagine. Thus, primitive dogs perform more like wolves on the rope pull test, making it more likely the modern domestic dogs' tendency to rely on humans (at least for those tested–other types of dogs may not) is an effect of selection subsequent to the separation of these dingoes from other dogs. The New Guinea dingo willingness to gaze into human faces is probably not a domestication effect, and if not, then this trait in domestic dogs is likely inherited from the dog ancestor.

FIGURE 5.1. *A female New Guinea dingo giving the photographer the steady "Singer Look."*

Dingoes as we know them, both the Australian and New Guinean subspecies, are wild dog populations that have evaded at least the last 5,000 years of selection for co-habiting with humans (minimum estimate by genetic dating of how long ago they arrived in Australia and New Guinea[7]). They are the closest we can come to what dogs may have been like before full domestication. Perhaps the dogs' greater propensity for looking directly into faces in a non-threatening manner is an ancestral trait, one that made it easier for dogs to associate with humans. People could relate to them through mutual eye contact, as they do with other people. The fact that the dingoes did not look at me during the blocked rope trials merely reflects their self-sufficient and self-confident "wild" temperament, rather than an unwillingness to make eye contact or ask for assistance. In other situations, the dingoes will actively solicit people to get them something they want that is out of their reach, such as a toy or food item, but when that something is in reach they apparently do not think of looking to the person for assistance. J. Topál, A. Kis and K. Oláh (2014, p. 330) summarized the general consensus of researchers on the domestication effect reasoning:

> Findings from dog-wolf comparative research indicate an early emerging and permanent preference for face-to-face interactions and eye contact with humans in dogs, but not in hand-reared wolves, and support the idea that this behaviour may reflect the dogs' adaptation to the human social environment.

MATCHING QUESTIONS TO APPROPRIATE SUBJECTS

The studies targeting individual dog and wolf behaviors are interesting, and as data accumulates perhaps valuable insights into dog and wolf minds will eventually be revealed. To get a better picture of current dog cognitive abilities, more types of dogs, including terrier and spitz breeds, need to be tested. However, unless the goal is to understand modern dog traits rather than "original" (before intense artificial selection) or ancestral dog traits, the subjects should be dingoes and aboriginal village dogs (see Part II). In one of the handful of studies that used Australian dingo subjects, B. P. Smith and C. A. Litchfield (2010) found the dingoes, unlike domestic dogs, had no problem with detouring around a see-through barrier to reach a reward. The detour task is intended to show spatial problem-solving abilities. The results: Dingoes were successful at figuring out they had to go around and did so in less time and with fewer errors than dogs tested in previous studies. In past research on detours, wolves had also done much better than dogs.[8] Smith and Litchfield concluded that their dingo results supported the idea that captive-raised wild canids are more adept at nonsocial problem solving than domestic dogs.

Extending the results of these tests to domestic dogs in general is questionable, as the dog subjects in all the studies have been the types of dogs specifically selected for hundreds to thousands of years for an increased ability to inhibit their behavior and accept human control. While it might be true, even probably true, that captive wild canids are better at nonsocial problem solving than the derived types of dogs—after all, the wild canids depend on this ability to survive—we cannot say this with any certainty because so far only a handful of wolves and Australian dingoes have been tested. Subject performance in behavior testing depends on the subject's personality traits, such as boldness/fearfulness, their past experience, the capacity of their senses, and the design of the tests. Variations in any of these factors (and more) can skew results. So, although this type of research is important, generalizing results to "all" of any species at this early stage of the science must be done carefully, with caveats. A. M. I. Auersperg, G. K. Gajdon and A. M. P. von Bayern (2011:141–144) say that "[W]hat is missing is . . . a battery of tests establishing species differences that might affect performance in different cognitive tests, such as object exploration, motivation, attentiveness and fear/phobia." They conclude:

The outcomes of Auersperg et al. [2011] illustrate how even diminutive differences in non-cognitive behavioral components such as neophilia or morphology can mask and/or interfere with the respective cognition involved and impact on the species' performance. It highlights how different performance[s] in problem solving task are not always symptomatic of species differences in cognitive ability or general intelligence. . . . In future comparative research, establishing behavioral and psychological profiles of the species to be compared ought to precede comparative tests of specific cognitive skills or general intelligence. This may help to identity problem solving tasks that are equivalently applicable to the target species and hence achieve a high degree of 'compatibility' of the obtained data.

J. Hecht and A. Horowitz (2015) point out that the practice of inbreeding to produce and maintain breeds of purebred dogs has resulted in specific behavioral tendencies being enhanced, and this needs to be taken into account when analyzing behavior. The tendencies are not due to changes in the underlying genetic programs for the behaviors, which are present in all dogs, but to variation in the "thresholds" of stimulation needed for the dogs to express a behavior. Greyhounds, for instance, have a very low threshold for chasing fast moving objects, and the livestock guarding breeds mentioned earlier have a very high threshold for chase. As more breeds are tested with standardized methods, results can be combined into over-all tendencies for specifically selected groups of breeds.

The accumulated results of the Canine Behavioral Assessment and Research Questionnaire (C-BARQ) developed by J. A. Serpell and colleagues at the Center for Interaction of Animals & Society at the University of Pennsylvania School of Veterinary Medicine have verified there are reliable differences between breeds on various traits, including (using their definitions) trainability, attention-seeking, excitability, and aggression. J. A. Serpell and D. L. Duffy (2014) report that, for example, retrievers tend to rank high on trainability while hounds rank low, and huskies rank low on attention-seeking, while dachshunds and toy poodles rank high. If these standardized test methods comparing breed and species traits are extended to comparing different types of dogs, that is, of functionally-based groups that have had similar long-term selection, such as gun dogs (pointers, retrievers, spaniels), sight hounds, scent hounds, terriers (all dogs developed to go to ground after prey), sled dogs, companions, and aboriginal land races, we will have useful, relevant information.

RESEARCH AND INDIVIDUAL DIFFERENCES

The study of the individual behavioral abilities of dogs and wolves is rapidly expanding, and wolf social behavior has been studied both in captivity and the wild for decades. However, very little serious study of dog social behavior has been done, except for their social interactions with humans. The main reason for this is that, until recently, there was little money available to study dogs. Then scientists realized dogs have many of the same diseases as humans. Once they started studying the genetics of the dog as a model for locating disease-related genes in humans, the funding opened up. Afterwards, data on dog mental and neurological problems started coming in, showing that neurological/mental illnesses in dogs may correspond to the same problems in people. This of course sparked an interest in the dog's mind and behavior.[9] It seems strange that until very recently we knew much more about the minds, abilities, and social behavior of monkeys (the former favorite surrogates for humans), pigeons, rats, and mice than we did about dogs. Dogs were difficult study subjects for the old style in-house research. They are much more expensive to keep than lab animals other than monkeys and apes, even the small beagles that have been the work horse of research laboratories, and have a slower reproduction rate that extends the time needed for generational studies. In the last couple decades of the 20th century, the public's objections to dogs being experimented on ended most of the research kennels, except for dog food companies. Now, with modern communication systems, it is relatively easy to recruit large numbers of dog owners to volunteer their pets as study subjects, or to gather information by using standardized owner reports as in the C-BARQ questionnaire mentioned above. Scientists can select from this volunteer pool dogs that are similar in life experience.[10] This new way of doing dog behavior research makes it possible for many more scientists to do affordable studies.

Research on inter-dog social behavior was slow to get started, probably because of the widely held assumption that dog behavior is essentially just artificially changed, or "defective," wolf behavior. Wolf studies were done because they are an iconic wild predator important to understand, and in most places a threatened or endangered species. The consensus was that we had the ancestral template from wolves to transfer onto dogs. Also, the impetus to study dog behavior was weak due to the reasonable feeling that, since we have lived intimately with them for many thousands of years, we know dogs well. In fact, the sense of familiarity, combined with the assumption wolves had already taught us what we needed to know about dog behavior, obscured for a long time the realization dogs are not merely altered wolves, and they deserve scientific study.

The most damaging (to human/dog relationships) of the dog-behavior-equals-wolf-behavior notions is actually based on a popular misconception about wolf behavior. This misconception is about how wolves develop and maintain social hierarchies within their packs, the idea that the wolf's natural social group is a rigidly structured pack, ruled by a dominant Alpha male and female pair using force and intimidation. Supposedly the Alpha or "dominant leader" male and female wolf are dictators, with the other same-sex pack members taking up successively lower positions in the pecking order.[11]

This misunderstanding of wolf pack behavior comes from an erroneous interpretation of how the Alpha leaders maintain relationships with other pack members, based on research by R. Schenkel (1947) on a captive group of zoo wolves. Schenkel reported a lot of aggressive encounters between pack members, and he

FIGURE 5.2. *Adolescent wolves greeting their elder pack member with strong submissive behaviors.*

thought the Alphas of both sexes forced submission of same-sex pack members. Typical of wolves on exhibit in zoos at the time, this was not a natural family group, but a group put together from unrelated wolves from other zoos. Decades of field research since then have revealed that wild wolf packs are normally just families, and the Alphas are the parents. As the oldest and most experienced individuals, the parents automatically become the leaders. Their offspring acknowledge this status by voluntarily submitting to them. There may be a lot of ferocious-sounding growling and snarling by parents and flashing of fangs toward offspring, but this is almost always merely ritualized threat behavior, communication with no intention to do harm. The Alphas do not "force" pack members to submit. The younger pack members willingly honor the parents' status by deferring to them. Their survival depends on cooperation and heeding their parents' corrections. The relationships between the remaining pack members are based on individual characteristics such as age (younger siblings subordinate to older) and personality.[12]

Here is how packs normally form. A lone dispersing male and female, usually from different packs, find each other and bond, then locate an area with adequate resources (water, enough prey to support a family, den sites, no wolf competitors) where they can establish their territory. This wandering stage can take more than a year. The following spring they produce a litter. Wolf offspring that survive to adulthood usually disperse in one to three years, depending upon the amount of preferred prey in the pack territory. If prey is abundant, offspring will stay with their parents through one or more breeding seasons, while they mature and learn to hunt. The older siblings benefit the parents by helping to feed, puppy-sit and entertain (play with) their younger siblings. The offspring benefit by practicing skills in relative safety. Eventually, competition for food within the pack may reach a critical point. The parents then start to limit the feeding of the older pups by driving them away from kills. The parents, especially the dam, need to be in the best condition for the next breeding season, and the growing pups still too young to fend for themselves must get enough nutrition for growth. Although rare instances of two litters being raised by one pack have been reported, in at least one case where a subordinate female produced a litter, the pups apparently died of malnutrition.[13] The mature pups eventually move off, wandering sometimes hundreds of miles while looking for a mate. Very rarely will they encounter a pack that will adopt them, unless it is one that has lost an Alpha and the newcomer is the same sex as that Alpha. The truth is, pack hierarchies are not formed from top down forcing, but created and maintained by the subordinate wolves through affiliative behaviors (behaviors which promote group cohesion using friendly/positive gestures) toward the parents, and of younger siblings toward the elder offspring.

ADAPTABILITY AND PACKS

Adaptability is the foremost characteristic of the social organization of most canid species. Whether canids live in groups, pairs, or singly typically depends on external factors such as resource availability, especially the size and number of prey animals. Like the protein coding nuclear DNA discussed in Chapter 4, the same inherited genes can underlie a range of behaviors. To recap, the gene regulating sections of non-protein-coding DNA are responsible for the same gene being expressed in different ways, producing variability in things like leg length and body size. For behavior, both variable gene expression and external environmental factors determine what behaviors are expressed. The how, where, when, and why of a behavior varies according to a multitude of internal (genetic, neurological, and physiological) and external (physical and social environment) variables. These variables govern which behaviors are expressed and what form they take.

One primary difference between natural wolf and dog social behavior has been obscured by the curtain of the wolf ancestor myth. The evidence is fairly clear that free-ranging dogs do not typically form wolf-like organized packs. Neither do feral dogs or dingoes, as far as is known. For dogs, the simple definition of "pack" by T. J. Daniels and M. Bekoff (1989) is the best fit: "A pack was defined as a group of animals that traveled, rested, foraged, and hunted together."[14] Dog pack hunting is less organized and efficient than wolf pack hunting. The interpretation is usually that this lack of organized packs is a domestication effect, that the ancestral behavior program for pack living has been disrupted by selection since wolf and man became commensal. However, there are alternative interpretations: The dog ancestor was not specifically adapted to pack hunting large game; and/or packs are not necessary in the dog ecological niche. If the dog ancestor was not the large gray wolf, *Canis lupus*, it probably did not characteristically live in groups larger than a pair with that year's offspring. In either case, wolf packs are not the rigid, always cooperative integrated groups often visualized.

The majority of canids do not form long-term stable groups. The typical social arrangement is a monogamous mated pair that defends a territory.15 The offspring disburse when the next litter is born, or not, according to the abundance of prey available in the parent's territory. If there is a lot of food, some pups stay for a year, practicing their hunting and puppy raising skills. The interesting question is, why do wolves regularly live in groups? A wolf pair is perfectly able to raise a litter without "helpers" and there are three reported cases of single wolves, two female and one male, that successfully raised a litter (the male's mate died when the pups were old enough to eat solid food).

FIGURE 5.3. *The gray wolf may not be the best model for dog behavior. This wolf is European.*

Having extra pack members does not, as previously supposed, increase the efficiency of hunting beyond a pair's. Observed wolf pack size ranges from a pair with no pups to 42, the average range being from 4 to 11. About half of the members of larger packs are adults over a year of age, the rest young-of-the-year. Pack members do not always do things together. During the spring when pups are too young to leave the den area, pack members often travel alone around the communal territory. Packs may also split and subsets travel separately for a period coming back together after days or weeks. Turnover of membership is high in wolf packs, with the only constant being the Alphas, at least until one or both die. If an Alpha dies, it may be replaced by a pack member or by an immigrant.[16]

The review of wolf social ecology by L. D. Mech and L. Boitani (2003a) proposes several reasons for wolves staying in groups. Wolves can take up to 5 years of age to become fully mature, and as long as there is plenty of food, it is in both the parents' and offspring's best biological interest to let the grown pups hang around. The offspring gain valuable knowledge and skills, and the longer association ensures the parents' investment in them is not wasted. The idea that hunting in groups increases efficiency as measured in food-per-wolf is wrong (but is true for the smaller coyote when they prey on deer[17]). Actually, as pack size goes up, less food is obtained per wolf. The reason wolves maintain packs seems to be that, while killing large prey does not require packs, sharing large prey allows packs to work. Maintaining packs depends on both prey size and abundance, and moderate numbers of large prey or high numbers of small prey (or carrion) can sustain larger groups.

One of the factors promoting larger groups is that without a pack to defend a large carcass, such as of a moose or elk, scavengers soon consume everything but the largest bones. The wolves would not get to come back to complete consuming their kills. Ravens associate with wolves and are often the first scavengers to arrive at a kill, daring to try for bites even when the wolves are resting nearby. In one observation 135 ravens were counted at a wolf-killed elk, and in another they removed up to 81 pounds (37 kg) from a fresh moose carcass.[18] Mech and Boitani (2003a:129) conclude that "By this analysis, a lone wolf might lose two-thirds of a moose kill to scavengers, and a pair of wolves half their kill, while a pack of ten wolves would only lose about 10%." Long-term cooperative packs make biological sense for most wolves most of the time, reducing food waste and increasing offspring chances for surviving long enough to reproduce themselves. Given the dependence of pack living on the influence of so many external factors related to the canid ability to vary social behavior, basing our assumptions about, and interpretations of, dog social behavior only on the wolf is unjustified. We *must* investigate dog behavior without the preconceptions inherent in assuming the wolf is the ancestral template.

FIGURE 5.4. *Wolves and three species of scavengers, coyote (barely visible on the right), grizzly bear and ravens, at an elk carcass in Yellowstone National Park.*

FIGURE 5.5. *A Border Collie "giving eye" while herding sheep, an inherited canid stalking behavior artificially selected to be easily disconnected from the natural culmination of attack.*

DOMESTICATION EFFECTS

Whatever the dog ancestor was, there is no doubt the behaviors of domestic dogs have been changed from the ancestral version. Modern domestic dog behavior is shaped from birth by the effects of artificial selection acting on their ancestors over the last 10,000+ years. Because of this selection, dogs are adapted to living in intimate association with and being dependent upon humans. Some dog behaviors have been changed from the "wild type" (probable generalized ancestral form) by artificial selection. For instance, the wild type canid chase or stalk that ends with a "kill bite" to kill or disable the prey has been fragmented in many dogs. While most domesticated dogs will chase and even pick up small animals, some have lost the instinctive bite necessary to feed themselves as predators, or at least it is not expressed if they are well fed. The behavioral elements of predation have become, for many domestic dogs, just a form of play. Chasing prey-like things still feels good to them (from the intrinsic reward from a flood of neurotransmitters in the brain's reward system), but the overwhelming urge to kill when the prey is caught has faded. In some dogs even the chase instinct has been reduced. Breeds developed for traditional sheep guarding, for example, will not leave their flock even when they see potential prey they could chase.

Other dog behaviors are exaggerated by special selection. One well-known case is the stalking posture combined with intent staring (called "giving eye") that Border Collies use to herd sheep. Stalking is of course a component of hunting in canids, which is why sheep are predisposed to move away from dogs performing this behavior. In the case of Border Collies, stalking usually does not result in attacking the sheep because the urge to complete the hunting behavior was selected against, while sensitivity to learning to follow human commands was increased. In some cases, dogs are incapable of certain behaviors due to the physical exaggerations and mutations of form imposed on their breeds by human selection. Long hanging ears and tightly curled or absent tails cannot stand up to send clear signals of social intent or emotion. All the above are examples of valid domestication effects on individual behaviors and morphology that effect behavior. Whether or not other aspects of dog behavior, especially social, that differ from wolf are effects of domestication is questionable and depend upon whether or not the wolf was the actual dog ancestor.

INHERITED BEHAVIOR PATTERNS

The behavior of complex animals is the result of both inheritance and learning, and the learned portion is often the main determinant shaping an individual's behavior. It is much harder to do scientifically valid behavior research with animals than with humans because they cannot tell us their past history or describe their feelings. Even humans are often unaware of what past experiences have shaped their behavior: The "why." Teasing apart the varying inherited and learned factors underlying animal behavior is often impossible. Determining with any certainty what part inheritance plays in the behaviors an animal performs would require its physical and social environment to be carefully controlled from conception (in-utero environment and nutrition can have a profound effect on behavior) to adulthood. This would be not only impractical but extremely expensive. In any case, behavior is the result of the coordination of many different body systems, and it is the "potential" for a behavior that is inherited. So ethologists (ethology is the study of natural behavior in the wild, or in species-appropriate settings, and views behavior as an evolutionarily adaptive trait) look for behaviors that are typically present in all individuals of a species, or in more than one closely related species, no matter what environment they have experienced. They reason that such universal behavior must have an inherited basis. Then they look for the explanations of how the behavior varies between individuals or species to determine in what way experience, learning, and environment affect the inherited behavior.[19]

There was a long-term (over 18 years) multi-canid-generational study at the University of Connecticut of the inheritance of a behavior present in coyotes and jackals but absent in dogs and wolves: The open mouth defensive threat gape. When they feel defensive coyotes display a wide open mouth gape-threat accompanied by an arched back posture, usually with a hissing vocalization and a toe-tip mincing gait. The domestic dog and the gray wolf typically exhibit a snarl-threat in defensive encounters, often accompanied by a growl vocalization, and with no back-arching. The facial expression of a gape-threat is a common behavior in many kinds of animals including primates, foxes and cats (both of which also arch their backs), and reptiles. The researchers bred a beagle to a coyote, and then bred various offspring together for three generations and subjected them to variable social circumstances to track the inheritance of the three main coyote display components: Gape-threat, arched back, and hiss vocalization. They found that the three components are inherited separately. They also discovered that hormone levels, specifically cortisol, which becomes elevated in stressful social encounters, affected the expression of the gape-threat. Higher cortisol levels "triggered" the display components. So, in the beagle-coyote hybrids, social behavior appeared to function as a triggering mechanism, regulating cortical hormone levels which determined which of an alternative set of genes were expressed as behaviors. Alice Moon-Fanelli (2011:869) concluded that:

> Thus, social behavior modifies physiology, which re-regulates gene expression, which in turn, modifies behavior. . . . There are two other interesting inferences from our results. Based on the similarities of dog and wolf threat behaviors that set them apart from other Canidae, as discussed above, we suggest that these behaviors serve as taxonomic indicators of their affinity. The behavioral data are in agreement with the morphological and genetic studies (Lawrence and Bossert 1967, 1969; Shaw 1975; Wayne 1993)."

FIGURE 5.6. *Coyotes (top) and black-backed jackals (below) have defensive gape threats. Wolves and dogs do not.*

DOG SOCIAL BEHAVIOR MODELS

Very little research has been done on the social behavior of natural dogs (free-ranging aboriginal village dogs, feral dogs, and wild Australian dingoes[20]). In what has been published, no long-term cooperatively hunting packs have been reported. Occasionally a group of dogs will reportedly kill a large animal, but only the Australian dingo does so on a regular basis. The difference between the domestic dogs and dingoes is the feral or free-ranging dogs are basically scavengers of human waste and not predators, while the dingo is fundamentally a predator that will scavenge when it can do so while avoiding humans.

NATURAL DOGS

Given the amount of artificial selection their ancestors have been subjected to, modern domestic dog breeds and mixed breeds, even feralized, are not the best subjects from which to determine natural (ancestral) dog behavior. The dingoes, on the other hand, have existed as predators, independent of human assistance, for at least 4,000 years (and in my opinion, probably more than 12,000 years) and are the best remaining models for natural dog behavior. Whether the dogs that founded the Australian and New Guinea dingo populations, discussed below, were somewhat domesticated or not is irrelevant. That far in the past, any selection would likely have been only for a reduced tendency to resort to physical aggression toward humans, without altering other behaviors. The fact that neither population of dingoes ever voluntarily adopted the niche of camp scavenger, the one all other free-ranging dogs fill today, is an important clue that their ancestors arrived with a full set of natural behaviors.[21]

The free-ranging, free-breeding ancient aboriginal village dogs are the best existing models for natural domestic dog behavior. L. Boitani, P. Ciucci and A. Ortolani (2007:152) support this by saying:

> Although data are limited, village dogs may be the most abundant category of dog in the world . . . and, so far, they have been the most overlooked of all canines. Yet they hold the key to understanding how a predator evolved into the most faithful human companion. Studying the behaviour of village dogs is of critical importance to shed light on the complexities of human-dog interactions in our society.

Called "landraces," these regionally-adapted dogs have been only slightly affected by artificial selection, mainly in the direction of less aggression (both toward humans and other animals) and increased self-control. With minimal training as puppies, village dogs learn they must coexist with poultry and livestock and are completely reliable thereafter. This would not be true of the dingoes that will be discussed later. Some aboriginal landraces are described in Part II.

Available evidence indicates aboriginal village dog (AVD) populations have existed

FIGURE 5.7. *An Africanis village dog ignoring poultry in Kwazulu Natal, South Africa.*

for many thousands of years in the rural temperate, subtropical, and tropical Old World. While there are still ancient landrace dogs in more industrially developed areas, especially in Asia, due to the cultural environment (for instance, dog keeping traditions), even the free-ranging dogs are owned, sheltered, and fed, their reproduction is often controlled, and puppies culled from litters for physical attributes such as size. This means increased artificial selection, which devalues these populations as far as being representative of more natural never-controlled aboriginal dog populations. In higher latitudes, the winters are too harsh to support free-ranging dogs year-round and any aboriginal dogs, such as Siberian laikas and Inuit dogs, are well controlled, including their reproduction. Recent genetic studies indicate the most rural populations of free-breeding aboriginal dogs are still pure landraces, not mixed with modern dogs from Europe and Asia. These populations are probably direct descendants of the first dogs to populate those areas.[22]

Free-ranging AVDs live only near human habitation, surviving on handouts, feces and garbage. This is the niche dogs have filled since they became attached to humans. Only a few have actual personal relationships with people. They are commensal but hardly any ever become what would be called pets, that is, fed and purposefully sheltered as part of a human social unit. Even fewer are trained to do any sort of tasks under human command, but those who have been trained performed at least as well as most modern purposefully-raised dogs.

The reigning King of Thailand, the honorable Bhumibol Adulyadej, admires the traditional village dogs of his country, which are very similar to the AVD populations of other areas. The Thai dog still is free-ranging and freely-breeding. Trying to promote respect for and the keeping of Thai AVDs because they are so well adapted to the local environment and a cultural relic, the King wrote a very popular book about his favorite AVD pet. He has also sponsored an annual festival where the public can see trained Thai dogs perform various tricks. The idea of actually training village dogs, that they are capable of learning commands, is novel to most traditional people who grew up seeing the dogs merely living around them scavenging.[23]

FIGURE 5.8.
A traditional Thai village dog.

As motorized transportation, and therefore the movements of people, have increased over the last 100 years or so, the village dog populations are becoming increasingly mixed with imported modern European and Asian dogs. Because the village dogs have always been "just there" and never appreciated by most locals as the highly adapted and intelligent dogs they are, the locals admire the developed imported breeds. Even a mixed-breed mutt is an irresistible exotic status symbol. As a result, populations of unmixed village dogs are sometimes just increasingly rare remnants. Perhaps efforts like those of the Thai king to reveal the intelligence and trainability of these dogs can encourage the people to value their village dogs.[24]

INDIAN NATIVE DOGS

Despite their continual presence for thousands of years in many subtropical and tropical rural areas of the Old World, the only in-depth long-term observations of aboriginal dog social behavior have been done by Sunil Pal on Indian AVDs.[25] His 1999 Ph.D. dissertation was on the ecology and behavior of the free-ranging dogs in West Bengal, India. Those working to secure some conservation status for the Indian village dogs prefer the name Indog, for Indian native dog for this AVD landrace, which will be used here. Pal continued his study on the same population for more than 10 years, through several dog generations. He reports that dogs did form what he called "groups" which maintained variably sized territories (the size dependent on the available resources in that area) by cooperatively expelling dogs trying to join from other groups, and that there was a social hierarchy within each group. In each group one female and one male received the most submissive gestures and initiated the vast majority of aggressive

FIGURE 5.9. *A group of Indogs in central India.*

interactions. They could be called the "alpha pair" in the sense they are the social center of the group. However, these groups are not the same as the tightly-knit wolf family pack that forages as a unit. Indogs form temporary bonds with mates and offspring, and like any collection of individual animals including humans, some have bolder personalities and become the natural leaders to whom the others defer. The groups Pal observed were stable over time, although there was some turnover of members due to death, immigration and emigration.

The Indog groups Pal studied ranged from 5 to 8 individuals and their group home ranges on the outskirts of a town of 4,200 inhabitants varied between 8.9 and 23.2 acres (3.6–9.6 ha). The larger areas were utilized during the breeding and whelping seasons. All of the hundreds of aggressive encounters observed within each Indog group consisted of only growling and snarling. Pal says "chasing and severe fighting were never observed" among group members (1998, p. 343). The highest rates of aggressive encounters, both within and between groups, occurred during the annual breeding season as males attempted to court females. The next highest rate was during the denning season when lactating females aggressively defended food and the immediate area around their dens from all but their mate. Although Pal (1998) describes the Indog alpha females as "highly aggressive" toward other females in their group during the seasonal breeding period, this harassment by the alpha did not prevent all the females of each group becoming pregnant each year. In Pal's (2005) study of parental care, all 6 bitches in the two groups observed became pregnant. In wolf packs the alpha female is usually successful in preventing the other females from breeding, as a way of assuring there will be enough food and assistance to raise her litter. Scavenging dogs do not require a pack to support a litter so there is no natural restriction on reproduction other than environmental (basically the available nutrition).

BREEDING AND REPRODUCTION PATTERNS

Despite the belief that dogs, unlike wild canids, are promiscuous breeders, the Indog females had strong mate preferences. The estrus alpha female of each group was aggressive toward non-preferred males when they attempted to court her, soundly rejecting them. A total of 36 adult males attempted to court the 6 females in the 2005 study, and yet 4 females who each had between 5–8 males courting them bred with only one male. Of the two other females, one mated with two males out of the 6 courting her, and although the other allowed all 4 courting males to mount, successful copulations were observed with only 2 males. Pal also reported that 3 of the females mated only with the same male in consecutive years, and that the majority of the mates chosen were not closely related. This demonstrates female Indogs have a fairly strong mate preference and that they retain the natural instinct to breed to genetically less related mates.[26] Males do not have the same biological imperative to choose the "best" mate, unlike females who have to invest so much time and energy in a litter. Instead, males invest energy in mating with as many females as possible, which increases the chance the genes of the successful will be perpetuated.

Villagers often allow Indogs to have their pups in sheds or under buildings. Bitches were not provisioned by their mates or other group members, and foraged daily for themselves. Pups are nursed for about 6 weeks. All females regurgitated for their pups between 6–11 weeks of age. Of the successful sires, only 4 stayed with the litters for the first 6–8 weeks. Three of these were the males that mated exclusively with the dam, and the fourth was one of the two to successfully breed the dam that season. These males defended the nest area when the dams were foraging, but their nest attendance declined after 5 weeks of age when the pups became mobile. One attending male started regurgitating for the pups when they were 9 weeks of age, and the dam then stopped doing so. He fed the pups for 10 days. Captive NGD males also guard their litter and regurgitate for the pups.[27] This disproves, at least for natural dogs, the commonly stated difference between male dog and male wolf behavior, that wolves provide parental care while dogs do not. Given the right circumstances where there is a social bond between the sire and dam and the sire has access to the dam and pups, perhaps more male domestic dogs would regurgitate for them and provide guard services for the pups. In both male and female canids regurgitation for puppies is partly reflexive, triggered by the behavior of begging pups licking and pawing at older dogs' mouths. Dams regurgitate with the least stimulation by puppies due to hormonal priming, but even some unrelated adolescent and adult dogs will feed puppies if highly stimulated. In the wild, usually the only older individuals interacting with pups would be family, and feeding close relatives is positively selected for because it helps ensure that a genetic line successful in that environment survives.

FIGURE 5.10.
Paternal care by the Indog subjects at den sites in S. Pal's study.
A. The male on the right is grooming a puppy.
B. A sire in S. Pal's study regurgitated.

FIGURE 5.10B.
Before regurgitation.

FIGURE 5.10C.
After regurgitation.

Indogs average 5–6 pups per litter. Pup survival is low, with only about 30% surviving to 3 months of age, when they become independent. In villages, pups are culled by people, many get killed by cars, and after they start exploring on their own at about 8 weeks of age, they are susceptible to predation by other dogs and wild predators.

AFRICAN VILLAGE DOGS

The African village dog of Zimbabwe is the only other free-ranging AVD systematically studied. J. R. A. Butler and colleagues did a series of surveys and studies to define the ecology of Zimbabwe's rural village dogs.[28] Unlike the Indogs, nearly all of the African village dogs (labeled "Africanis" by those struggling to preserve them) were owned with someone claiming that dog as theirs, and the dogs showing fidelity to a particular home.

However, they were not confined, nor often purposefully fed, and scavenged nearly all of their food. Dogs in several rural areas were surveyed for type, age and size. The percentage of dogs judged to be pure AVDs varied from 85.5% to 99.2%, with the highest percent of mongrels in areas closest (but still distant) to urban areas.[29] Average size was about 33 lbs (15 kg) and 18.6 in (47.3cm) shoulder height, so the African AVDs in that area are about the same size as Indogs. The landscape in the Butler study area is typical for communal lands in Zimbabwe (land shared by members of an indigenous community) – a mosaic of agricultural fields interspersed with fallow fields, regenerating woodland, and woodland. Homesteads are groups of huts surrounded by unfenced yards. Garbage is tossed into open pits near the yard boundaries. The people keep free-ranging mixed herds of cows, goats, donkeys, and sheep that wander in the harvested fields and the woodlands. Some keep free-ranging chickens *(see Figure 5.7)*.

In the areas the Butler group surveyed, more than 60% of the households owned dogs. The main reason given for keeping dogs was for guarding against human intrusion, although village dogs probably would not guard in the usual sense of defend. They merely serve as a warning system. The second most frequent reason for keeping dogs is guarding crops against baboons and other crop pests, and the third was guarding livestock and poultry. Only 1.2% of owners listed pet as a reason for keeping a dog. The main study was done in a settlement adjacent to a wildlife reserve, to determine if dogs were killing game or competing with hyenas, lions and leopards. They radio-collared 16 adult dogs that lived within 1 mile (1.5 km) of the reserve boundary fence, then observed them and the other dogs in the area for about 7 months. They discovered the dogs did not go farther than about 1 km (0.5 mile) into the reserve. Only one larger male dog (weight 49.6 lbs/22.5kg) was successful in hunting large ungulates (2 impala, 1 kudu), which he learned to run into the fence. J. R. A. Butler (1998) reported that other than those larger species killed by that particular male, the 10 wild prey killed by dogs during the study were all small to very small, < 2 lbs. (1 kg) to 19.8 lbs. (9 kg). Dogs also killed six goats and one sheep. Most of the observed kills were by two dogs. The larger prey was killed inefficiently, by repeated mauling body bites, which Butler called "their clumsy manner of killing vertebrate prey" (Butler, 1998, p. 89). This is contrasted with the behavior of dingoes and other wild canids and cats, which often kill with throat bites *(See Figure 5.11)*.

The diet of the African AVDs, by weight, was 87% human-derived food (sadza – a gruel of sorghum and maize, the staple of the people – vegetables, domestic fruit, domestic animals, human-caught fish, and human feces). The rest of the diet mainly consisted of scavenged wild mammals, springhare (a rodent that weighs about 6 pounds [2.7 – 3.6 kg]), insects, grass, and wild fruits. Only 13.3% of the total consumed was deliberately fed, the rest scavenged. The available human toilets were open-pit, but most people merely defecated in the bushes. Feces contributed 25.3% of the diet, with human feces making up 23.6% of that. Domestic animal remains, mostly goat and cow but also dog, were available as carcasses of animals that died naturally were left in the

FIGURE 5.11. *A golden jackal with a fawn it killed with a throat bite.*

open for the dogs and vultures to clean up.

The dogs were judged to be in good body condition, and the females had no trouble reproducing so the diet was more than adequate. The carrion and human feces supplied most of the protein. Carrion accounted for between 8 and 43 g/100g protein and human feces about 18.7g/100g. In many areas of rural India the dog diet is probably much the same as in Africa.

The African AVDs were active about 25–30% of every 24 hours, moving about in a leisurely search for widely spread, small food items. Home ranges of the radio-collared subjects were all centered on their homes, but each included at least one other homestead that was home to other dogs. Home range size varied from 83 acres (33.5 ha) to 358 acres (145.0 ha), much larger than in India where the human population, and therefore waste, was more concentrated. The African dogs' living area was a settlement of small holdings (individual properties with a house) and not as in India a "village" per se, which usually also contained small businesses selling food and meat, and so had more concentrated AVD food sources.

Similar to the Indogs, the African AVDs showed only rare dog-dog aggression above the threat level. Home ranges were not defended against other dogs, but they did expel non-resident dogs from their immediate homestead area. The survival rate to the adolescent stage was also about the same, 30%. Fifty percent of pups died in the first month, despite plentiful food, mostly due to disease.

FIGURE 5.12A. *An African village dog dam with pups in Kwalazulu Natal, South Africa.*

FIGURE 5.12B. *A pair of Africanis breeding in the Eastern Cape region, South Africa. Note the relaxed, nonaggressive postures of the males waiting on the sidelines.*

THE DINGO AS A FUNCTIONAL MODEL

The dingoes provide the best model of possible dog ancestral behavior because they have been independent of human support for at least 4,000 years. From what is known about dingo social behavior, in many aspects it is more like coyote or jackal behavior than wolf. The dingo, coyote and jackal are predators adapted to taking mostly small to medium-size prey (from mouse size to about dingo body size). They do not form stable long-term organized packs like wolves because their body size and life-style do not require packs. While single wolves and pairs can survive capturing prey the size of deer and can manage for a time in areas where there is a super abundance of small prey like rabbits, their body size and heavy skulls and mandibles (jaw bones) are designed for hunting very large prey, many times their individual body size. For wolves, small prey like rabbits, hares or mice are sufficient only as a temporary stop-gap substitute during times when large prey is scarce. Using that large wolf body to capture such relatively tiny prey consumes almost as much energy as the prey provides, unless the prey population is so dense the area is saturated and no traveling to search for them needed. Therefore, wolves naturally prefer, for instance, elk and deer to small game. Even if small prey are available, wolves will choose the larger prey until those species are so rare locating them takes too much energy. Small size prey can support dingoes comfortably even during reproduction.

The reason wolves cannot depend on small prey is that populations of small animals often follow boom-and-bust cycles of a few years.[30] The population increases until they are so common that it takes little effort to catch plenty of them, so they can provide enough food for wolves. But when the small mammal population outgrows the resources to support it, it declines rapidly. After the population crash, the natural resources – food and places to raise young – are again abundant for the survivors, and the population starts to increase. When the population of small prey is low and scattered over large areas, it takes as much energy for a wolf to hunt them as it gets from eating them. Propelling that large wolf body around searching for and chasing small animals uses up about as much energy as capturing large prey, but with much less nutrient return. Once they evolved large body size and the related constraint of living in the long-term packs that can efficiently use large prey, wolves could not "de-evolve" back to a significantly smaller body size. Therefore, in general terms, the wolf is an obligate pack predator and almost strictly a carnivore.

Dingoes, jackals and coyotes are "facultative" pack predators. They can make a living as singles, pairs, or family groups (packs) depending on the size and density of the available prey. For example, coyotes hunt in small family groups where deer are the main prey species available. But where they depend on rabbits and other small game, the mated pairs raise their litter, which then disburses before the next litter is born rather than sticking around. Coyotes and jackals are also facultative carnivores that do not depend solely on meat. They, at least intermittently, consume significant amounts of non-animal food such as fruit.[31] The smaller more omnivorous canids have larger grinding surfaces on their molar teeth than wolves, an adaptation to eating vegetable matter. As mentioned in Chapter 2, dogs have grinding surfaces slightly larger than the wolf, closely matching those of the coyote and jackal, but combined with canines that

FIGURE 5.13. *The Asiatic golden jackal may be a better model for dog social behavior than the gray wolf. These specimens were photographed in India.*

significantly differ from other *Canis* species, adding to the dog's niche singularity.

When it originally occurred to me that it was unlikely the wolf could be the direct ancestor of the dog, as a behaviorist my first idea was to get the most primitive type of dog and study its behavior. I reasoned that the most primitive dogs "should" have the most wolf-like behavior if they descended from the wolf only 15,000 – 40,000 BP. I was thinking about Australian dingoes (ADs), which 25 years ago I knew only from some documentaries and a few sections in dog books. At that point, very little research had been done on wild dingoes except how to kill them (predator control), but they existed as wild predators. Because they had not been subjected to artificial selection like domestic dogs, their behavior would be as close to the ancestral type as it is possible to find today. I quickly learned that only zoos could get ADs out of Australia and there were only a handful in American zoos.

CAPTIVE NEW GUINEA DINGO BEHAVIOR

Searching through dog books for an alternative study subject, I came across the New Guinea singing dog in B. Wilcox and C. Walkowicz's 1995 book *The Atlas of Dog Breeds*. While their common name seemed a bit silly, the entry said they are a wild dog and very rare in captivity. There was a picture of some of these dogs residing in North American zoos. They looked like smaller, shorter-legged ADs. Intrigued, I wrote to the authors asking where I could learn more, and they gave me Dr. I. Lehr Brisbin, Jr.'s contact information. The North American captive population consisted of just a handful of specimens. Dr. Brisbin was trying to get some interest in them from the scientific community, while attempting to keep the captive population going by locating people who would breed them responsibly. Some scientists questioned if these were merely aboriginal village dogs, but from the first DNA results they closely matched the AD, not AVDs or modern dogs, so the correct name for them is New Guinea dingo (NGD), although Singing Dog is still the most popular common name. The AD and NGD are geographically isolated wild subpopulations of dog, and therefore are considered subspecies, not "breeds" of dog.[32]

Reading what little was available on these dogs, and a little investigation into the NGD's natural habitat in the remote uninhabited mountains of New Guinea, convinced me they were indeed another ancient line of dog that had a population still existing as wild predators.[33] A year and a half later, Dr. Brisbin sent me my first two NGD puppies, littermate females Buna and Tufi. A few weeks later I was able to buy a male, Kai, from a local exotic animal dealer. Within a few minutes of getting 8 week old Buna and Tufi home from the airport, I realized that these dogs were different – really different – than both the domestic dogs and the captive wild canids I had experience with. At that time, I already had 25 years of professional experience training domestic dogs, hundreds of them, and had bred three breeds, plus worked hands-on with captive coyotes, wolves and wolf hybrids. The NGDs were my awakening that dogs are more unique than most have assumed.

The behavior of the captive NGD is very different from the wolf. Nothing is known about the behavior of the wild NGDs because they live in extremely remote and rugged mountains and have not been studied. In captivity, mature adults can only be kept in opposite-sex pairs due to intense same-sex aggression during breeding season. Two of the same sex will fight to severe injury or death over a potential mate because in this subspecies the normal submissive postures that cut off aggression in wolves and most canids are not effective. In other words, if the attacked NGD throws itself down and rolls over, the most extreme submissive posture in canids, the attacking NGD does not "honor" this signal and continues to attack. Captive parent NGDs allowed to raise their pups begin to aggressively harass them when the next breeding season is about to start, and the stressed offspring must be removed. This same-sex adult aggression and rejection of same-sex adolescent offspring are indicators that NGDs in the wild cannot maintain long-term groups.[34] Since in their natural mountain habitat, about 8,000 feet (2,500 m) and above, there are only small size (0.5 – 3 lbs. [0.2 – 1.25 kg], compared to NGD size of 19 – 32 lbs. [8.6 – 14.4 kg]), widely scattered prey, permanent packs would not be adaptive. Therefore, NGD pairs undoubtedly maintain defended territories, excluding other NGDs including their own adult offspring.

Until wild NGDs are studied, all that can be said is some captive NGD behavior

Chapter 5: NATURAL DOG BEHAVIOR?

differs significantly from wolf behavior. For instance, the typical NGD howl is unique, not wolf-like; furthermore, NGDs are neophilic like most dogs (attracted to investigate novel things) while wolves are neophobic (fearful of novel things).[35] Unlike wolves, even with the abundant food available in captivity, NGD parents aggressively reject same-sex offspring when their next breeding season starts. Wild NGD howls have not yet been recorded, but it is extremely likely the wild and captive howls will match. It will be interesting to see if wild NGDs are as same-sex aggressive after maturity as the captives. Of course, in the wild the offspring can avoid parental aggression by keeping their distance, unlike the captives. Since the captive NGDs, until recently only kept in zoos, have not been directly selected for any characteristic, their behavior is probably still "natural" and representative of wild NGDs. C. Künzl, et al. (2003) found that wild cavies (guinea pigs) show no behavioral differences after 30 generations of captive breeding, but the domesticated guinea pig has a reduced physiological reaction to stress and is less exploratory. They concluded (Künzl, et al. 2003, p. 187): "These data suggest that the long-term breeding and rearing of wild guinea pigs in captivity do not result in significant changes in behavior and hormonal stress responses. It appears to take much longer periods of time and artificial selection by humans to bring about characters of domestication in wild animals."

FIGURE 5.14. *Perching on objects is a universal canid behavior. The New Guinea dingoes provide an illustration of a subspecies specific behavior unique only in frequency. They perch at an extremely high rate compared to domestic dogs, strongly preferring to rest on elevated surfaces. These subadult NGDs would soon be separated to prevent fights.*

WILD AUSTRALIAN DINGO BEHAVIOR

Unlike wild NGDs, wild ADs (hereafter just "dingo") have been studied, and can serve as a model for some natural wild dog behavior. Unfortunately, the early studies were to understand dingo predation on livestock and the effectiveness of control efforts (thus the reason for funding studies), and most recently to understand dingo's possible effect on the survival of endangered species. I could find no long-term observations dedicated to the social life of an individual family group. We do know that, like jackals and coyotes, dingoes are facultative pack hunters. The most commonly sighted social group is a pair or dam with or without the current year's offspring. In all areas that have them, dingoes, which are 29–40 lbs. (13 to 20 kg), prefer to take the larger kangaroos and wallaroos, which vary by sex and species between 35 lbs. (16 kg) – 190 lbs. (85 kg). These larger species require pack hunting, so groups of two or more work together. In areas where kangaroos are scarce or absent, dingoes survive readily on small prey such as water fowl, wallabies, possums, and rabbits, hunting singly or in pairs. P. C. Thomson (1992: 543) reported: "The increased utilization of smaller prey by dingoes coincided with changes in sociality (disintegration of packs and an increased number of solitary dingoes). . . . Dingoes cooperating in groups were more successful than solitary dingoes in hunting large prey (kangaroos, calves)." The dingo's flexible social organization allows versatility in hunting strategies and adequate defense of resources. Free-living dingoes consume about 7% of their body weight daily. A rabbit averages about 2.4 lbs. (1 kg), the perfect size for a single dingo's daily ration. Since dingoes fairly easily capture rabbits, they are the most valuable food for dingoes today.[36]

FIGURE 5.15. *Wild dingoes scavenging at the refuse dump of an isolated mining station in the Tanami Desert, Australia.*

Like all canids and many other animals, dingoes will scavenge from carcasses and utilize garbage when it is available.[37] T. Newsome et al. (2014a) studied the dingoes of the western portion of the Tanami Desert over an area covering approximately 11,185 mi2 (18,000 km2). Because the dumps at the two mining camp study sites are the only anthropogenic sources of food in an otherwise uninhabited area of desert, the amount of scavenged food and what prey the scavenging dingoes ate could be directly compared to that of desert dingoes who did not scavenge. Results of analysis of the 1,907 fecal samples collected from non-scavenging dingoes showed small mammals had the highest probability of occurring in feces (note this is a simple presence count, not a percentage of volume) of 63.1%, while reptiles had a 51% probability, and invertebrates (insects) 32.8%. The scavenging dingoes' feces contained rubbish more than 60% of the time, but also contained the other categories indicating that they were not "dependent" on the garbage. The results also showed clear prey preferences for the two dingo populations. At the mine sites, they consumed mammals at a higher rate than reptiles year round, while the non-scavenging dingoes switched to eating more reptiles during the warmer season when reptiles are more available. Thus the scavenging dingoes' natural seasonal switching to the most abundant and easy to catch prey was interrupted, with possible negative local consequences for continual predation on the area's small mammals. One noteworthy finding was the scavenging dingoes showed no resource defense, allowing free access to the dump for transient dingoes. Since the food supply was superabundant they had no need to waste energy protecting it. It would be interesting to know if these "transient" dingoes permitted to forage were formerly associated with the regular scavengers, or were perhaps even related, but that information was not available.

Dingoes are now living in suburban areas near large cities. B. Allen et al. (2013) state that dingoes are present in most large cities and towns in Australia, but I could find no DNA sequencing reported for "urban dingoes," and they could be dingo/domestic dog hybrids. Allen et al. (2013) radio collared (GPS) nine subjects, seven of them juveniles, and monitored them from 5 to 43 successive days between October 2005 and June 2006 in the Pine Rivers Shire and Maroochy Shire of sub-tropical southeast Queensland, where over a million people live. Unfortunately, no social interactions (two or more dingo collars indicating they were in close proximity) were reported. All subjects were active at dusk and through the night, traveling 6–9 mi (10–15 km) during the night within their home range, then significantly decreasing their activity within a few hours after dawn. These dingoes probably are more nocturnal than usual in order to avoid the times of greatest human activity. They visited dumps and traveled through back yards. The adult collared male was much more active during the day than the other subjects. His experience may have made him more confident around human activity. He and the adult female used the urban habitat much more than any of the juveniles. All juvenile and sub-adult dingoes monitored spent no more than 19% of the time in urban areas, while the two adult dingoes spent 62–72% of their time in urban areas. All the dingoes frequented the densely vegetated bushland and caneland habitats within the area, probably because more prey was available there and the brush provided shelter during the day. Temporary dens/resting places were located in brush within 219 yards (200 m) of houses *(See Figure 5.16)*. Each dingo had a varying pattern of habitat use. Allen et al. (2013, p. 134) point out this is a very preliminary study, especially given there were only two adult subjects, and conclude: "It might reasonably be assumed that observed movement patterns and activity levels would reflect the foraging strategies and nature of risks an individual urban dingo is prepared to face, and variable results for different individuals might be expected for such a highly adaptable species."

DINGO BREEDING

From reports to date, all adult female dingoes are bred during the annual season, which varies according to latitude and weather (breeding seasons are delayed a few weeks during droughts). Except in times of high prey availability, despite most females becoming pregnant, each dingo group raises an average of one litter a year. A few instances of two litters successfully raised have been seen. The oldest and youngest females (9 – 10 mos. old at first breeding season) often do not produce litters despite being bred. Dingo pups remain at den sites (plural because the dam may move them from the natal den to others) and nurse for about 8 weeks. For the next month they travel around with the dam, who leaves them in places with shelter while she hunts, then feeds them by regurgitation. Although there are anecdotal reports of alloparental care (care provided by an individual other than the biological parent) in wild dingoes, there is only one published direct observation report of any wild non-dam feeding pups, an adult male of unknown relationship to the pups, who regurgitated for a litter.

FIGURE 5.16A. *Given suitable soil and time to excavate, dingoes can construct elaborate dens. This one was dug in a large enclosure by captive dingoes and had several entrances and tunnels. Here a dingo enters the den.* **FIGURE 5.16B.** *A dingo exits the almost-invisible entrance.*

After about 12 weeks of age the dam brings the pups pieces of prey or whole small prey. By 4 months old they accompany adults (not always the dam) on hunts, and by 6 months they are capable of independence, although most associate with adults of their familiar group for at least several more months. There are periods of stability in which dingo groups maintain an unchanging mosaic of almost exclusive territories, and there is little emigration. But, dingo groups break up during droughts when prey populations decline, and also when their own numbers reach an unsustainable level given the local prey density. The original territory may be retained by remaining dingoes, portions divided up among near-by groups, or it can be appropriated by immigrating dingoes from other groups.[38]

FREE-RANGING FERAL AND DOMESTIC DOG SOCIAL BEHAVIOR

If stray and feral dogs, despite their different genetic and experience backgrounds, have some social behaviors that are similar to that of the dingoes and AVDs, this could be an indication those behaviors are species-typical for all dogs. There have been a few studies on stray dogs, mostly in urban settings, but as these are obviously owned, or previously owned, dogs that are merely not confined, and their past experience is unknown, there little useful information to apply to the question of natural dog behavior.[39] However, there are three exceptions which are true also for dingoes. First, stray dogs were most frequently sighted alone and the median number of dogs sighted together was always around 1.5, even in areas with high populations of free-ranging dogs. Second, while they had home ranges–a defined area that they spent most of their time in – they were not aggressively territorial. Third, like the beagles studied by F. A. Beach and B. J. LeBoeuf (1967) and the Indogs, females showed some mate preferences, rejecting some males and favoring others. B. Ghosh, D. K. Choudhuri and S. Pal (1984) found that Indog females did not prefer large, aggressive males. After reviewing stray dog studies, S. Spotte (2012:165) concludes: "In summary, both the extent of a male's familiarity with the estrous female and female choice – not rank in any dominance hierarchy or body size relative to the other males – are important factors influencing mating success in dogs."

T. J. Daniels and M. Bekoff (1989) found that both the urban and rural free-ranging dogs in their study were predominately solitary, and the overall survival rate of puppies to 4 months of age was 34% for the litters of five females. In other studies of urban free-ranging dogs they were also mainly sighted alone, and pup survival past weaning was about the same, one-third. Daniels and Bekoff (1989) suggest that similar factors may influence social organization in all of the sites where free-ranging dogs were studied. Perhaps it is more efficient for single urban dogs to scavenge from trash bins. But, still, when they are satiated, it is curious they do not seek each other out just for the company. Maybe they do rest someplace together when not scavenging and this was not observed or not included in the published accounts. It could be that fully domesticated dogs may not be as strongly socially inclined as has been assumed based on the wolf model, or the dogs observed wandering were actually owned and their need for social interaction was fulfilled at their homes.

FERAL DOG MODELS IN ITALY

There have been three studies of feral dogs that are relevant to natural dog behavior. From 1981-1983 D. W. Macdonald and G. M. Carr observed and radio-collared free-ranging dogs in two villages and feral dogs in one adjacent large grassland surrounded by forest in the Abruzzo mountain region of central Italy. The grassland area dogs were again observed and radio tracked by L. Boitani, F. Francisci and G. Andreoli intermittently from 1984 to 1987. The third study, by a group of researchers, was of feral dogs outside of Rome, Italy, in a 741-acre (300 ha) reserve of grassland and wooded areas. To simplify referring to these multi-author sets of studies, I will refer to them as the "Abruzzo" and the "Rome" studies.[40]

The Abruzzo grassland (meadow) site contained a large communal garbage dump and two smaller single village dumps, which provided all the food for the feral group and occasionally for stray dogs from the nearest village. The feral group did not show aggression or competitive exclusion toward unfamiliar dogs encountered at the dump (where food was super abundant) or near villages, but did defend with cooperative barking and threatening body language if strange dogs crossed into the core areas of their territory. Interestingly, Boitani et al. (1995) reported reduced aggressive tendencies of group members during estrus periods, when strange dogs often arrived, not the increased incidence of aggressive interactions as reported for AVDs. In fact, three dogs (one male, two females) that arrived during a group member's estrus period were recruited into the group.

The main feral Abruzzo dog group size (two smaller feral groups were not directly studied) varied over the years of the study from 3 to 15, depending on the number of juveniles present. The Abruzzo dogs were larger than long-term feral populations studied in other places, being between 37-68 lbs. (17-31 kg). This is a sheep-producing area, and subjects were undoubtedly mostly mixed versions of the Abruzzo mountain sheep guarding dog. Few lone individuals were sighted. Most sightings were of 2–6 individuals. The group was decimated by poison baiting in the autumn of 1983, leaving only one male who became the leader of a newly formed group recruited from immigrants, the subjects of the 1984–1987 study. Seven adult dogs were poisoned during the study, so the group structure was frequently in flux. Even so, the core of the group when it was stable consisted of two pairs that maintained their relationships until one of them was killed or disappeared (fate unknown). Unlike the village dogs also studied during the Macdonald and Carr (1995) study (this part not discussed here), which were most frequently sighted alone, the feral dogs were generally seen in company of certain

FIGURE 5.17. *Abruzzo mountain dogs*

other group members, mostly in pairs but also in regular subsets. The exceptions were when bitches left the group to whelp and den with their pups.

Natal dens were all within 109–218 yards (100–200 m) of the group's regular resting sites, except one dam who denned 9 miles (15 km) from the usual group center of activity, near a dump seldom used by the study group. Although group members visited the dens on occasion, none were seen providing direct care to the dam or pups, and the dams raised pups on their own until the few survivors joined the group after 8–10 weeks of age. However, in a separate small group consisting of a pair that produced two litters during the study period, the male stayed closely associated with the female after pups were born. He slept in or near the den and played with the pups when they were mobile, although he was not observed feeding the dam or pups. L. Boitani et al. (1995) reported 40 pups born alive, with 28 (70%) dying within 70 days from birth, 9 (22.5%) living to 120 days, and 1 (2.5%) dying before one year of age. Only 2 pups survived beyond a year.

There were two packs of wolves in the area of the Abruzzo studies.[41] The dogs' main territory was in between those of the wolf packs, with the dog group home range overlapping each of the wolf territories by about 30%. The wolves of this area of depleted large natural prey obtain much of their food from scavenging human waste, just like the dogs. They are smaller than most Holarctic gray wolves (but larger than the dogs), do not form stable multi-generation packs, and frequently are sighted alone. One of the wolf packs regularly utilized one of the village dumps in the study area as a food source, but they found no evidence of direct wolf/dog interaction there. However, a juvenile dog was killed by wolves near another dump, probably because he was caught traveling alone.

The Italian group's activity budget was: Resting 48% of the time, active 40%, and traveling 12%. Nursing dams had greatly reduced activity and travel levels, staying at the den and making only one trip per day to feed at the dump. Overall, there were significant activity peaks at dusk and dawn. The dogs laid up in the higher elevation forested areas during the day, traveled in morning and evening, and foraged at dumps at night, probably to avoid human activity. L. Boitani et al. (1995, p. 231) describe the fundamental configuration of the group activity as: "The dogs show a basic pattern of becoming active at their resting sites around sunset, travelling to the dumps, staying active around the dump areas and at lower altitudes [of their home range], and then traveling back at dawn to their resting sites." They mention that the dogs likely scavenged at night to avoid humans, but that this does not explain the bimodal (dawn/dusk) activity pattern reported for all studies of dogs in all seasons (e.g., Scott and Causey 1973; Daniels 1983a,b; Daniels and Bekoff 1989a). Around 50% of canid species are typically nocturnal, but some that normally have other inherent activity patterns learn to forage at night around human habitation. J. Aschoff (1966) suggested circadian rhythms with regular peak activity levels is an innate trait. Coyotes and golden jackals, the closest canids is size to natural dogs, are active at all times of day but peak at dusk/dawn. They also become almost exclusively nocturnal when foraging in human-dominated landscapes. It seems dogs may be the same.

The Rome studies were of a feral group of 25 to 40 free-ranging dogs living in a large nature reserve area on the outskirts of Rome, Italy. Some had undoubtedly been born "in the wild" and others were strays that chose to live farther from humans. At the start of the study, people had been providing dog feeding stations and water daily at regular places along the road for 10 years. Most of the food was waste from slaughter houses. During the study three dogs died, two disappeared, and six dispersed, mostly into other feral groups. Fourteen dogs of the original study group were still present at the end of the study. Unlike the other feral dog studies to date, these concentrated on the inter-individual behaviors of the subjects, especially those related to social rank and reproduction.

Like the Indog studies by S. Pal and colleagues (but unlike the stray dog studies), the Rome research found there was a stable social hierarchy in the group. Because dogs were individually identified, the social interactions among the dogs could be quantified. They recorded the number of times each individual was successful in initiating activities, and how often an individual was the receiver of submissive "friendly" or affiliative gestures (like low tail wagging and muzzle licking, mostly during greetings) and agonistic submissions (avoiding eye contact, lowering the head, lowering the tail between the hind legs, lying down on the back or yelping), in response to their aggressive displays. The statistical tests of significance on the inter-individual direction of observed submissions revealed a significant social hierarchy. The oldest individuals were at the top, with various inter-personal relationships among the lower ranking individuals. Leadership status was determined by which individuals were the most successful in initiating group movements and activity shifts from resting to traveling; from resting to feeding or drinking; from resting in an area exposed to the sun to a shaded area; and, from resting in an open area to resting in the dense vegetation. Except for traveling to feeding sites, movement was over relatively short distances. There was no single over-all leader, but a few older individuals, who received the most formal submissive greetings from the other group members, habitually initiated activity. These behaviors were interpreted as the acceptance of the dominance relationship by subordinates.

The high-ranking dogs that received friendly greetings from several group members were more likely to lead during group departures than "dominant" dogs that received submissive gestures only in aggressive/agonistic contexts. More individuals also strongly preferred resting near habitual leaders rather than other group members. The collective movements of the group seemed to arise because the subordinates wanted to maintain close proximity to specific social partners. Aggressive behaviors were rarely displayed between group members in the absence of sources of direct competition (food), and no contact fights were observed. So, in dogs as in wolves, dominance postures (upright head and tail, stiff body, direct stare), along with the voluntary submission of subordinates, are sufficient to maintain a stable social status; age affects dominance relationships in all contexts.

R. Bonanni and S. Cafazzo (2014) of the Rome studies reported that dogs in the study group were seen moving with one or more companions in 71.68% of sightings, and only rarely sighted alone. The 50–80% solitary dog data reported in most other free-ranging and stray dog studies has been used as evidence for lack of group structure in dogs. Bonanni and Cafazzo point out those studies were based on much shorter periods of observation, but the main difference was they were not specifically designed to collect data on social behavior. Pal did find some social structure in Indogs, in that individuals had preferred companions and mates, and they collectively defended their home range from unknown dogs. The only other studies on truly feral dogs living in a large group, the Abruzzo, also reported only infrequent single individual sightings.

WHY DOGS FORM GROUPS

In villages and suburban areas, food is usually available in small amounts distributed around the home range, so it is consumed immediately without need of a group to defend it. Individual resident domestic and village dog pets are almost always successful in repelling intruders from their small home area using threats, and smaller groups of AVDs easily defend their home range from somewhat larger intruding groups. Dogs who knowingly trespass are at a psychological disadvantage, which is why small domestic dogs can repel much larger dogs from their home yards using body language and vocal threats. Why, then, do at least some feral dogs live in groups larger than a pair? Macdonald and Carr (1995) felt that the main reason the Abruzzo dogs tended to travel as a pack was their likelihood of encounters with wolves. The dogs rarely traveled into wolf territories, and the only dog killed by wolves inside a wolf group's home range was one assumed to have been traveling alone. In group encounters between canids (dingoes, free-ranging dogs, foxes, wolves) at a shared, localized resource (food, water), the numerically smaller group invariably relinquishes the resource to the larger group when it arrives, withdrawing without direct contact. Often, the smaller group is repelled by the larger group barking or howling from a distance.[42] The Abruzzo wolves foraged singly or in pairs and would avoid foraging at the dump when the dog group was feeding. The wolves did not trespass into the core area of the dogs' home range, so apparently the dog group was large enough to exclude wolves. The Rome group was, due to its size, able to secure a very large exclusive area that included rich food sources, water, and woods for shelter during resting and denning, unchallenged by the other smaller groups in the general area.

According to the Abruzzo research, high ranking females benefited from the group because they achieved a greater rate of reproduction as judged by survival of offspring to adulthood, probably because they denned in the group's core area where few if any predators hunted (cats, badgers, foxes, jackals, wolves, other dogs). Lower-ranked females whelped farther away where the pups were more vulnerable to predation. S. Cafazzo et al. (2014) reported another reproductive benefit of being older and higher ranking. They were preferred mates. Estrus females spent more time in the proximity of high-ranking males who displayed friendly behavior towards them and were more likely to reject males that attempted to intimidate them. Males made more courting effort toward higher-ranking females. There are also possible, but not proven, benefits for the lower-ranked group members. Bonanni et al. (2010) point out that, as in wolves, the lower-ranked dogs were generally younger than the leaders, so it is possible that they benefit from their time being followers through development of their social or foraging skills.

R. Bonanni and S. Cafazzo (2014:89) concluded dog social behavior has been affected by domestication:

> To summarize this part, our belief is that ecological pressures experienced by dogs in the domestic environment have led to the evolution of a promiscuous mating system characterized by a reduction in cooperative breeding relative to wolves, and of a feeding strategy primarily based on scavenging and predation of small- or medium-sized animals, still involving some degree of cooperative hunting.

However, R. Bonanni and S. Cafazzo (2014:95) also suggest that the similarity of free-ranging dog behavior, regardless of different genetic backgrounds in different environments, comes from a common dog ancestor, without specifying that ancestor: We stress . . . that some of the salient features of dog sociality that emerged in our and in other studies of free-ranging dogs (e.g. dominance, cooperation) appear to be shared by multiple breeds and also by the Indian dogs [Indogs], suggesting a possible derivation by a common ancestor.

SUMMARY

The results of the studies discussed above indicate that all free-ranging dogs studied to date have the following general traits in common:

- In all climates, dogs have a bi-modal activity pattern, being the most active in the mornings and evenings.
- Dogs from fully domesticated populations are not typically predators. Even when feral, they are scavengers dependent on humans. The exceptions are the dingoes.
- True aggression (physical fighting) among dogs is rare. Avoidance of strange dogs and maintenance of social distance through visual threats and barking is the norm.
- Unlike a wolf pack, for example, free-ranging strays and village dogs form loose groups of familiar interacting individuals within a defined home range, typically traveling and scavenging individually or in pairs.
- Feral dog groups have leaders whose status is maintained by the other individuals voluntarily deferring to them and preferring their company, not by threats or aggression from the higher ranking dogs.
- All adult females in groups can and do breed; there is no social control of breeding.
- While there is no social control of reproduction in dog groups, both males and females show some mate preferences.
- Feral dogs and dingoes typically do not dig natal dens with entrance tunnels and a central chamber like those reported for wolves and coyotes. Instead, dens are usually located in natural cavities (in rock outcroppings, hollow logs, etc.— See an exception in Figure 5.16), or the enlarged burrows of ground-dwelling animals. Self-dug dens are often shallow holes, without entrance tunnels.
- Dams raise pups on their own with little or no direct support from other dogs.
- Pups become independent (able to survive on their own) between 4–6 months of age.
- In all environments, pup survival to age of independence is about 30%, and to age of reproduction about 10%. Dingo pup survival rate to reproduction is unknown.
- In all populations of long-term free-ranging, freely-breeding dogs, individuals tend to be between 25 – 40 lbs. (11.3–18 kg), with single (smooth) or double (inner insulating under fur and longer outer hair) coats according to the local climate, and to have upright ears and fish-hook shaped tails. Smaller or larger body size, and long or curly hair are severely selected against in the scavenger niche.

The fact is, the Holarctic gray wolf is an anomaly in the *Canis* species group, what is called an "outlier" as far as size and social behavior, and therefore the least likely model for the ancestral dog. While the small amount of currently available behavior data does not confirm that the ancestor of the dog was "another wolf" other than the gray wolf, one that typically did not hunt in groups, it does strongly suggest this possibility. There are three wolf species, the Arabian, Indian, and North African—all with body sizes nearer to aboriginal dogs and dingoes than to the Holarctic gray wolf—that typically do not maintain multi-generation packs. J. W. S. Bradshaw and H. M. R. Nott (1995) hinted at this when discussing the lack of ordered social hierarchies among companion dogs and the variability of dog behavior, calling for more study that treats the dog as if it is a distinct species without using the wolf as a template.

The best examples of natural dog behavior are disappearing. The remaining unique populations of the dingoes and the ancient aboriginal dog races are becoming increasingly mixed with modern dogs. If we want to truly understand our companion dogs, the social behavior of these remnants of the natural dog must be carefully studied in their native habitats before they are gone. The aboriginal dogs and dingoes are equally separated in time from the ancestral dog as modern breed dogs, but have been subjected to much less artificial selection for behavior traits. *If a behavior is predominant in all populations of aboriginal dogs despite their adaptions to various local conditions, there is a high probability that behavior is ancestral.* Of all the natural dogs, the social behavior of the dingoes is the least studied, with no data currently available for wild Australian dingoes, other than proximity measures from GPS location collars, and not even that for New Guinea dingoes. The dingo subspecies were founded after the human-dog relationship began, by dogs descended directly from those of South East Asia (the population that seems to be the ancient ancestral genetic source for modern dogs), where even today the village dogs are not subjected to direct artificial selection. Since their introduction the dingoes have thrived as predators completely independent from humans, the only remaining truly wild dogs. Therefore, the social behavior of the dingoes would be the least changed from the ancestor's. If we want to determine how the dog, underneath all the changes we have imposed on it, truly differs from the wolf, to get to the true essence of "dogness," it can only be through studying the natural dogs.

NOTES

1. Here are some recent books which review our current knowledge about wolf and dog social behavior: Jensen, ed., 2007; J. Serpell, ed., 1995; J. Kaminski and S. Marshall-Pescini, eds., 2014; S. Spotte, 2012; B. Hare and V. Woods, 2013; Á. Miklósi, 2007.

2. Frank, H. and M. G. Frank, 1982; Frank, H., 2011; Kubinyi, E., Z. Virányi and Á. Miklósi, 2007; Scott, J. P. and J. L. Fuller, 1965; Scott, J. P., J. H. Shepard and J. Werboff, 1967; Udell, M. A. R., et al., 2014.

3. Freedman, D. G., J. A. King and O. Elliot, 1961; Klinghammer, E. and P. Goodman, 1987; Scott, J. P. and J. L. Fuller, 1965.

4. Fentress, J. C. 1967.

5. Gácsi, M., et al., 2009a, 2009b; Hare, B., et al., 2002; Hare, B., et al., 2010; Udell, M. A., N. R. Dorey and C. D. Wynne, 2008; Virányi, Z., et al., 2008; Wobber, V., et al., 2009.

6. Frank, H., 2011; Frank, H. and M. G. Frank, 1982; Frank, H. and M. G. Frank, 1983.

7. Oskarsson, M.C.R., et al., 2011; Sacks, B. N., et al., 2013; Savolainen P., et al., 2004.

8. Frank, H. and M. Frank, 1987; Mersmann, D., et al. 2011; Pongrácz, P., et al., 2001.

9. Khanna, C., et al., 2006; Parker, H. G., A. L. Shearin and E. A. Ostrander, 2010; Sutter, N. B., and E. A. Ostrander, 2004; Tang, R., et al., 2014; Tsilioni, I., et al., 2014; Vermeire, S., et al., 2012; Wayne, R. K. and E. A. Ostrander, 2007.

10. Hsu, Y. and J. A. Serpell, 2003; Pongrácz, P., et al., 2001. Stewart, L., et al., 2015.

11. Bradshaw, J. W., E. J. Blackwell and R. A. Casey, 2009; Van Kerkhove, W., 2004.
    This "dominant pack leader" idea was based on the now disproven assumption that the wolf's, and therefore the dog's, typical behavior is to form packs with linear, transitive social hierarchies (A is always dominant to B who is always dominant to C, and so on), and the pack structure is maintained by aggressive interactions among pack members. Promoted by a self-proclaimed expert dog trainer who became famous because of a television show, the use of intimidation and physical force (which may get rapid suppression of the unwanted behavior but causes questionable long-term changes) to train dogs and maintain social status over them is not merely incorrect, but damaging to human/dog relationships. The rigid set of "be a dominant pack leader rules" owners are supposed to follow, such as never allow your dog to go through doors before you, to go ahead of you on walks, or to get on your bed, and to not respond when your dog invites play, create conflict for no useful reason and prevent people from enjoying a close relationship with their dogs. Applied animal behaviorist Suzanne Hetts devoted a few pages of her 2014 book to debunking the dominant leader style of training.

12. Mech, L. D., 2000; Mech, L. D. and L. Boitani, 2003b.

13. Mech, L. D. and L. Boitani, 2003b; Packard, J. M., 2003; Peterson, R. O., J. D. Woolington and T. N. Bailey, 1984.

14. Because of the wolf connotation associated with the term "pack," here the more

relaxed association of familiar, interactive dogs will be termed a "group." Laurie Corbett (1995:95) suggested the term "tribe" for groups of wild Australian dingoes: "Radio-tracking studies in several regions of central Australia have confirmed that apparently 'solitary' animals are loosely bonded in small amicable groups (tribes) that have a single living area but tend to hunt separately."

15. See D. W. Macdonald and C. Sillero-Zubiri, eds., 2004 for descriptions of canid species' social organization.

16. Boyd, D. K., & Jimenez, M. D., 1994; Mech, L. D. and L. Boitani, 2003a, 2003b; Packard, J. M., 2003; Peterson, R. O., J. D. Woolington and T. N. Bailey, 1984.

17. Bowen, W. D., 1981.

18. Peterson, R. O., 1977; Stahler, D. R., 2002.

19. Bekoff, M., 1972; Ewer, R. F., 2013; Lehner, P. N., 1998.

20. See Glossary for definitions of the terms used for dogs as intended in this book.

21. Cahir, F. D. and I. Clark, 2013; Crowther, M. S., et al., 2014; Gollan, K., 1982; Johnson, C. N. and S. Wroe, 2003; Meggitt, M.J., 1965; O'Neill, A., 2002; Oskarsson, M.C.R., et al., 2011; Purcell. B., 2010; Sacks, B. N., et al., 2013; Smith, B. and C. A. Litchfield, 2009.

22. Boyko, A. R., et al., 2009.

23. Ardalan, A., et al., 2011; Brown, S. K., et al., 2011; Savolainen P., et al., 2002; Zeuner, F. E., 1963.

24. The landrace descriptions in Part II were written by people dedicated to gaining conservation attention for aboriginal dogs.

25. Pal, S.K., 2003, 2008, 2010; Pal, S. K., B. Ghosh and S. Roy, 1998b.

26. Pal, S. K., B. Ghosh and S. Roy, 1999; Pal, S. K., B. Ghosh and S. Roy, 1998a; Pal, S. K., 2005. The domestic dog females (beagles) tested in an artificial choice experiment by F. A. Beach and B. J. LeBoeuf (1967) also showed clear preferences for certain males.

27. Koler-Matznick, J. and M. Stinner, 2011.

28. Butler, J. R. A., 1998; Butler, J. R. A. and J. T. du Toit, 2002; Butler, J. R. A., J. T. Du Toit and J. Bingham, 2004.

29. Boyko, A. R., et al., 2009.

30. Korpimäki, E. and C. J. Krebs, 1996. According to R. Woodroffe, et al. (2007), the African wild dog (AWD) which normally takes prey over 100% of an individual AWD body size (average weight 55 lbs./25 kg. so similar to the body weight of the Indian wolf) does support packs with abundant smaller prey. Fifteen known AWD packs, in areas with depleted larger prey, were able to maintain their packs long-term by preying mostly on dikdiks, a small antelope (average 8 lbs./3.5 kg or about

15% of the AWD body size). Still, the dikdiks were twice as large in relation to AWD body size as rabbits and hares are to gray wolves.

31. Bekoff, M. and M. C. Wells, 1980; Bekoff, M., & Gese, E. M., 2003.

32. Mayr, E., 1970; Mayr, E. & Ashlock, P. D., 1991.

33. Brisbin, I. L. Jr, et al., 1994; Troughton, E., 1957; Troughton, E., 1971; Voth, I., 1988.

34. Koler-Matznick, J., et al., 2003; Koler-Matznick, J., I. L. Brisbin, Jr. and M. Feinstein, 2001.

35. Kaulfuß, P. and D. S. Mills, 2008; Koler-Matznick, J., I. L. Brisbin, Jr. and M. Feinstein, 2001; Musiani, M., et al., 2003.

36. Corbett, L., 1995; Corbett, L. K. and A. Newsome, 1987; O'Neill, A., 2002; Newsome A. E., P. C. Catling and L. K. Corbett, 1983; Purcell, B., 2010; Thomson, P. C., 1992a; Vernes, K., A. Dennis and J. Winter, 2001; Whitehouse, S. J. O., 1977.

37. Reports of omnivores and predators scavenging at refuse dumps are common. Given the opportunity, most mammals and birds will utilize free food when it is available. The following canids are frequently named: wolves, foxes (several species), golden jackals, coyotes, and of course dogs.

38. Catling, P. C., L. K. Corbett and A. E. Newsome, 1992; Corbett, L. K., 1995; Thomson, P. C. 1992c; Purcell, B., 2010.

39. Beck, A., 1973; Berman, M. and I. Dunbar, 1983; Daniels, T. J., 1983a, 1983b; Daniels, T. J. and M. Bekoff, 1989a; Font, E., 1987; Fox, M. W., A. M. Beck and E. Blackman, 1975; Majumder, S. S., et al., 2013; Majumder, S. S., A. Chatterjee and A. Bhadra, 2014.

40. "Abruzzo" studies: Boitani, L., et al., 1995; Boitani, L., and P. Ciucci, 1995; Macdonald, D. W. and G. M. Carr, 1995. "Rome" studies: Bonanni, R., et al., 2010; Bonanni, R., et al., 2011; Bonanni, R. and S. Cafazzo, 2014; Cafazzo, S., et al., 2010; Cafazzo S, et al., 2014

41. Boitani, L., 1983.

42. Bonanni, R., et al., 2011; Corbett, L.K., 1995; Mech, L. D. and L. Boitani, 2003b.

CHAPTER 6. THE NATURAL SPECIES HYPOTHESIS

When someone admits one idea and rejects another which is equally in accordance with the appearances, it is clear that he has quitted all physical explanation and descended into myth.
— *Epicurus*. Letter to Pythocles, 87. (Trans. R. W. Sharples)

The idea that the dog is a domesticated gray wolf is a hypothesis, without enough support to be a scientific theory.[1] As revealed in the first three Chapters, there are some pretty big holes in this hypothesis, and especially some unrealistic assumptions about human and wolf behavior. The physical and behavioral traits of dogs have repeatedly been analyzed under the hypothesis that the dog is an artificially created descendant of the wolf. By this hypothesis, the dog is not a species evolved under natural selection in Nature, it is a gray wolf changed from its natural state either by human intervention or due to living in a human-created environment. The purpose of the studies that openly support a gray wolf origin is specific: To explain how that wolf became the dog due to domestication, either through adaptation for scavenging human waste or by direct artificial selection. This preconceived bias precludes looking at results objectively, and it results in conclusions being interpreted from that bias. As discussed in previous Chapters, some of the resulting explanations concerning the dog and wolf differences were based on additional assumptions unsupported by even indirect evidence.

Here is my working hypothesis about the origin of the dog as a natural species:

The dog originated as a natural species, another descendant of the ancestor of the wolf. The dog and wolf are sister species. This ancestor was not adapted to hunting large game as a pack like the wolf, but was a generalist predator and scavenger more omnivorous than the wolf, one that regularly singly took larger prey than other canids of similar size.

Basically this hypothesis says that the dog was a "dog" before it hooked up with modern humans. Like any good hypothesis, this idea is subject to falsification (can be proved incorrect), is supported by several types of secondary evidence, and at least could be tested through ancient DNA or other means. Much of the indirect evidence, like the differences between dogs and wolves, and why wolves are not likely to have been tolerated by Stone Age people, was presented in Chapters 1–3. Chapter 5 revealed how little we know about the behavior of dogs under natural conditions, and that what is known indicates the dog's social behavior is more like that of other canid species than like the wolf. The rest of this Chapter will discuss other relevant points.

Currently, there is nothing directly conflicting with the Natural Species Hypothesis, so it meets the criteria for being "scientific." Tomorrow they may discover a series of wolf-to-dog shaped skulls from Paleolithic sites that confirms the dog did come directly from the wolf, proving my hypothesis incorrect. Then at least we would know for sure, and I would be happy even though my hypothesis was wrong. Hypotheses that are invalidated contribute to scientific progress by stimulating research and redirecting it to other possibilities.

Here are the basic points of my hypothesis:
1. The dog was never *Canis lupus*. It was a naturally evolved species before it attached itself to humans.
2. The skull and jaw shape of the dog indicate its ancestor was adapted to singly take prey half its own body weight or larger, and to cracking bones heavier than jackals can handle.
3. The ancestral dog was more omnivorous than the wolf (consumed more plant matter and insects).
4. The dog was originally adapted to forests where it hunted mostly by ambush.
5. The dog had the most common canid social system, that of mated territorial pairs, not the rare system of structured packs like wolves.
6. The dog and modern man met sometime around 50,000 BP as *Homo sapiens* was moving into Asia/SE Asia (the most speculative aspect).

As pointed out in the introduction, I am not the first to think that the dog could be a natural species. It has been mentioned many times in the last two centuries, most notably by T. Studer (1901), F.E. Zeuner (1963), H. Epstein (1971), M.W. Fox (in a 1973 article for the American Kennel Club Gazette and later in his 1978 book), and C. Manwell and C.M.A. Baker (1983). These authors speculated that the original dog may have been something like the dingoes and aboriginal landrace dogs, but they did not follow up the idea by examining the specific differences between dogs and wolves. B. Lawrence (1967), a zooarchaeologist who studied early dogs, did not consider the Holarctic gray wolf a suitable candidate for domestication, but mentioned the possibility of the extinct *Canis variabilis* or, as many others did because it was a smaller wolf and resembles aboriginal dogs, the Indian wolf, as potential ancestors. As it turns out, additional study of Indian wolf mtDNA shows they are probably not close genetically to dogs and gray wolves. C. variabilis died out at least 200,000 years ago so could not be a direct dog ancestor. But I think Lawrence was correct in her general questioning of the largest living *Canis* species being a likely dog ancestor.

NO FOSSIL HISTORY

The main criticism of this natural dog species hypothesis is that there is no fossil record of a mid-size canid with a relatively broad muzzle and high forehead before dogs started showing up in archaeological sites 14,000 BP. However, there are many living species with no or very incomplete fossil records, including chimpanzees, Ethiopian wolves and polar bears. All of the available specimens of purported human ancestors would fit in a small closet. So many factors influence the fossilization process, the discovery of fossils, and how they are evaluated, that absence of a record for some species' ancestor cannot be taken as proof of its non-existence (as the old saying goes: 'Absence of proof is not proof of absence').[2] Fossils of the original dog may be lying in a museum storage box somewhere, waiting to be recognized, as happened with the large Paleolithic short-faced wolves which were originally thought to be common gray wolves. The next hunt for Pleistocene fossils could discover the ancestral dog. But even if it is never located, the ancestor's characteristics can be tentatively deduced from ancient dog skeletons and living primitive dogs, as long as the bias that 'the dog is nothing but an artificially altered gray wolf' is suspended.

Extrapolation of the appearance and behavior of extinct species using similar living species as models is, if done objectively, a legitimate approach to get at least an approximation of what the extinct species was probably like. It is a common technique even in genetics, where they extrapolate from modern DNA to determine the ancestral DNA sequences.

There were many small to medium size wolf-like canids in the early and middle Pleistocene in Eurasia (2,588,000 to 126,700 BP, which corresponds to the Paleolithic era in archaeology). The earliest wolf-line canids were much smaller than the later species. For instance, B. Kurtén, a mammal paleontologist, says (1968:109): "Wolf-like animals of about the size of a sheepdog are found all through the Villafrachian (Etouaires) and seem to favour woodland faunas... but absent in the extreme steppe associations.... Perhaps ecologically it resembled the Indian wolf of the present day." The "extreme steppe" is where *Canis lupus* later evolved to hunt mega fauna. Because he was Finnish, the "sheepdog" he would have been most familiar with was probably the Finnish Lapphund, which averages 18 to 20 in. (46 to 52 cm) shoulder height and 37 to 42 lb. (17 to 19 kg). This is within the range of the Late Paleolithic dogs (no direct connection intended; just attesting to the existence of many early varieties of small *Canis* species). The Villafranchian period ended about one million years ago, so none of the canids living then was either a dog or gray wolf. These early wolf-like canids were very widespread in Eurasia, the Middle East and North Africa until about 100,000 BP. The direct ancestor of the living Holarctic, South Asian, and African wolves, and the dog, therefore is some medium-sized, extinct species. The wolf species we have today are mere remnants of what was a much greater diversity of species in the past, most of which were no more than about half the size of the living gray wolf.[3]

THE NATURAL SPECIES MODEL

Figure 6.1 is a hypothetical model depicting evolutionary trajectories for the dog and wolf, ending in one species of dog represented by the domestic dog and the wild dingoes, and the few wolf species that have so far survived extinction.

While it may never be certain which wolf-like species was ancestral to the gray

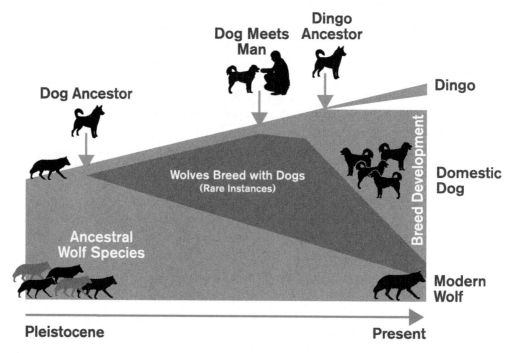

FIGURE 6.1. *The Natural Species Model for the origin of the dog. This model is intended to illustrate some general relationships of time and population sizes for the dog and wolf (not to scale).* **A.** *Left bottom: Pleistocene ancestral wolves. There were many more species of wolf in early to mid-Pleistocene than today. Most were smaller than modern wolves.* **B.** *Upper left: The dog's small wolf ancestor separates from other wolves by mid to late Pleistocene.* **C.** *Mid upper section: The dog/human relationship starts when dogs attach themselves to humans as scavengers, perhaps 40,000 BP. The commensal dog population started increasing rapidly about 15,000 BP, near the oldest estimated time the ancestor of the dingoes arrived in Australia and New Guinea. Far right: Purebred dog breed development began only a few hundred years ago. The domestic dog population today is many times greater than that of wolves.*

wolf and dog, according to the very complete comparative canid paleontology monograph published in 2009 by R. H. Tedford, X. Wang and B. E. Taylor, the extinct small wolf *Canis chihliensis* of Eurasia shares more characteristics with the larger *C. lupus*, and therefore the dog, than any other studied extinct canid. While the gray wolf evolved on the steppe tundra before the latest Ice Age, the dog could have been evolving from the same ancestor in the forests farther south, filling a different niche.

If this natural wild dog hypothesis is considered a possibility, what might this ancestral dog have been like? The only hints we can get are from the dingoes, the ancient aboriginal dog races and the prehistoric dogs of Europe and the Middle East. These are all about 25–45 pounds (11–20 kg). The dingoes live as wild predators on mostly small to medium size game and the aboriginal dogs as scavengers around human habitations. They have the body size that can best survive in those niches, and according to comparisons of modern village dog and dingo skulls with ancient skulls, they have not changed in the last 5,000 years.[4] As explained in Chapter 3, the original canid domesticated could have been larger than the later domesticants, and as domestication progressed the proto-domestic dog population adapted to the reduced nutrition available scavenging off of humans, becoming smaller. But, for reasons discussed below, I still think the original dog was no larger than about 45 pounds (20 kg).

My concept is that the dog filled a niche (the position or function of an organism within a community of plants and animals) between the large-pack canids like wolves, dholes and hyenas that together can take prey many times their body weight, and the jackals that are adapted to singly taking prey smaller than their own body size. This "in-between" niche is not filled by any living wild canid.[5] The moderate-size dholes specialize as pack hunters of prey much larger than their individual body size, but normally not as large as typical wolf prey. Therefore, dholes have strong jaws and teeth. The jackals do not have the skull or teeth for heavy-duty gripping of large live prey or consuming larger bones. When there is strong competition for the same resource in the occupied niches, similar species will try other ways of making a living to avoid the competition. In my hypothesis the dog ancestor avoided direct competition with the dholes and jackals, which evolved elsewhere and expanded into South East Asia (my hypothetical area of dog origin), by adapting to the vacant niche of individuals utilizing moderate size prey. Even within a single species, when the usual resources are being utilized near the maximum sustainable rate, some individuals will switch to another resource, such as prey of a different size. If they survive, those individuals become a viable population, which slowly becomes better adapted to filling the vacant

FIGURE 6.2. *The New Guinea dingo is smaller than the Australian dingo. This young captive male is about 30 lbs. (13.6 kg), a size comparable to the majority of prehistoric dogs and some early wolf-like canids. Dingoes are sexually dimorphic, with females usually 10–20% smaller.*

niche. Eventually, due to ecological factors the two populations diverge genetically and stop interbreeding. This is called "ecological speciation" and is one way biodiversity is developed and maintained.[6]

THE DOG NICHE

Modern domestic dogs gone feral are always mainly scavengers of human waste. None have been reported surviving solely as predators except a small population of dogs that went feral on an uninhabited Galapagos island and survived in a very harsh environment by hunting marine iguanas, at least until the authorities killed them all to protect the iguanas.[7] Some think that the dingoes are predators rather than scavengers like feral domestic dogs because in Australia and New Guinea they are not in competition with other wild canids; it is competition which supposedly restricts domestic dogs to living near humans. Perhaps this is true for modern domestic dogs, which have lost the coordinated suite of instincts to compete as predators in the wild with today's wolves, jackals and coyotes. However, it would not have been true for the ancestral dog. If the ancestor filled a different niche than the pack canids, there would be no direct competition and therefore no reason it would be out-competed by larger pack canids. The dog could have targeted prey not preferred by dholes or wolves, used a different habitat (such as forests rather than open areas; each habitat requires a different hunting strategy for both canids and humans), or partitioned resource use by being active at a different times of day than wolves, e.g., full day instead of crepuscular hours. The fully domesticated modern dogs have, over thousands of years, lost the ability to be independent of humans, leaving only the dingoes as natural predators. Some also think that the dingoes are able to thrive in the wild only because there are no large cats or other big carnivores, such as bears and hyenas, present that could prey on them. But golden jackals, a canid similar in size and niche to the dingoes, are still extant in Africa and SE Asia where there are (or recently were) both large and medium-size wild cats and dholes. In Africa, three species of jackal live where there are lions, leopards, African wild dogs, and spotted hyenas. If jackals can co-exist with larger predators and pack hunting canids, there is no obvious reason why wild dogs would not be able to do so, now and in the past.[8]

According to the most recent dating estimates from their genes, the founders of the Australian and New Guinea dingo populations came from SE Asia, arriving between 4,600–18,300 years ago.[9] I think it may have been much earlier, concurrent with the original human migrations to those areas around 50,000 BP, but that is not supported by any objective evidence. Whenever they arrived, the dingoes still had the intact instincts to remain wild predators. Knowing dingo behavior, I have no doubt if a breeding group were released into Eurasia or North America they could compete with other canids of similar size such as jackals and coyotes, and like those two species successfully adjust to the competition from wolves, while avoiding being killed by other larger carnivores.

FIGURE 6.3. *In Africa, black-backed jackals manage to feed at the same time as spotted hyenas, despite the danger to themselves. Perhaps the jackals wait until the hyenas are satiated and have a reduced urge to protect their kill, like the ravens that scavenge wolf kills.*

The Australian dingo population is becoming more hybridized with domestic dogs. Genetic evidence indicates in some areas Australian dingoes are now up to 80% hybridized. There is no stopping this genetic flow unless large reserves are set up for pure dingoes—large enough to sustain the dingoes in the long term as predators of natural prey. While this may be desirable, it is not practical. The only way to prevent domestic dog genes from continuing to enter the Australian dingo gene pool would be a dog and dingo-proof fence around the entire reserve.[10] The New Guinea dingo has so far been protected from interbreeding with domestic dogs by its very remote and rugged habitat.

The North American coyote is an example of a mid-sized canid coexisting with wolves. Studies in Yellowstone Park and other areas where wolves have returned or been reintroduced show the resident coyote numbers went down significantly, and then as they adjusted to a smaller niche stabilized at a lower number than when wolves were absent. In some areas where there is abundance of both large and very large prey and wolves, for example in Riding Mountain National Park in Manitoba, Canada, the coyotes make most of their living scavenging wolf kills. Deer, moose and elk are abundant in the park, and the presence of wolves does not affect the coyote population. When the coyotes there hunt their own large prey, they take deer, which in Manitoba is not the favored prey for the wolves. Wolves do opportunistically kill coyotes when they can, mostly when a coyote gets too bold trying to scavenge from fresh kills the wolves are still protecting. Whether or not large and mid-size canids can share the same territory, and how well each does, depends on the prey base.[11] What we know of canid natural history indicates the predator guild species (a group of species that exploit the same resources, usually in the same way) can adapt to shared territory if there is abundant prey, if they take different size prey, or if they hunt at different times. Even if both pack hunting canids such as dholes and small prey specialists (e.g., jackals) were present, there is no reason the dog would not fit into the late Pleistocene canid guild.

THE ANCESTOR QUESTION

It is a fair question to ask "If the dog was originally a wild species, what happened to this ancestral dog?" The ancestor is apparently extinct in the wild. Many ancestors of domestic animals have gone extinct or nearly so. The last auroch, ancestor of the cow domesticated about 10,000–8,000 BC, was killed off in Poland in 1627. The reasons aurochs went extinct seem obvious: They were large, their natural habitat was broken up and taken over by agriculture, and they were easily hunted by humans, who considered them a danger to people and competitors with the domestic stock. Eventually those that remained had no mates available. The horse was domesticated only about 4,000 BC. The wild horse known as the Tarpan, thought to be the closest relative if not the direct ancestor of the domestic horse, became extinct in the late 19th century. Przewalski's horse is the one remaining true wild horse species (as opposed to "wild" feral domestic horses). Barely hanging on now, it was saved from imminent extinction by captive breeding of a handful of specimens, and then reintroduced to its natural Eurasian steppe habitat. However, Przewalski's horse is not a direct ancestor of the domestic horse. At the end of the ice ages, horse populations dwindled as their steppe habitat was replaced with forests. Horses were also heavily hunted by humans, especially after domestic horses were available to round them up. Eventually the direct Eurasian ancestor of the domestic horse was wiped out. The wild Bactrian (two-humped) camel has been reduced to a very small population, estimated to be less than 600 individuals, in a few remote areas of Mongolia and China. Today the only "wild" dromedary (one-humped) camels are actually feral domesticants.[12]

On the other hand, the red jungle fowl, the main if not sole ancestor of the domestic chicken, still exists in the wild over much of tropical India and South

FIGURE 6.4. *This superficially appears to be a pure male Indian red jungle fowl, the main ancestor of the domestic chicken. However, this free-ranging specimen may actually have one or more domesticated ancestors*

East Asia. It is estimated to have been domesticated about 6,000 BC. However, jungle fowl have been breeding to feral domestic chickens as human populations introduce the domestic variety further into wild chicken habitat. The wild red jungle fowl's genetic integrity, like that of the Australian dingo, is being compromised by introgression from a domesticated form. Someday they may well be completely swamped by the domesticated genetic systems. Jungle fowl are highly neophobic and react almost instantly with escape flight when startled by movement, while domestic chickens have generally lost this escape behavior. The hybrids may not be able to survive in the wild without this behavior.[13]

From limited reports of recent mammal surveys of the wildlife reserves in South East Asia (SEA), where in my opinion the dog most likely originated, dogs not associated with humans are not present in the central uninhabited areas, so it seems there are no wild dogs left. Dogs do roam in the edges of reserves and manage to kill some prey, but shelter and reproduce around the villages.[14] Perhaps something like the jungle fowl scenario happened to the wild dog ancestor in SEA. If, as I suspect, the forest-adapted dog ancestor lived in only one rather restricted geographic area, then as domestication progressed the proto-domestic population (and later the rapidly-expanding fully domesticated population) surrounded the territory of wild dogs. Interbreeding continued until in the modern area there are only village dogs dependent upon humans. The domestic dog has lost the necessary suite of coordinated instincts to survive as a wild predator. After the domestication of other species, the last 5,000 – 9,000 years depending on the species, domestic dogs that could not curb their instinct to kill were eliminated from the gene pool. The result is that while domestic dogs retain the urge to chase and will attack animals, as adults their ability to catch and kill wild animals is in general awkward (with individual and specific breed exceptions), equivalent to the undeveloped ability of juvenile wild canids. Physically, the SEA village dogs may still generally resemble the ancestral dog. L. Corbett included SE Asian village dogs in his designation of "dingo" due to their similar appearance, but without any direct proof they are genetically the same.[15] The SEA aboriginal dogs still have the "dingo-like" ancestral form because they have not been selected for appearance, only for non-aggression toward humans and reduced predatory behavior (See Figure 6.5). Also, they have remained in the physical SEA environment where I believe the wolf-like ancestor evolved into the dog. Village dogs of other geographic areas have physically adapted to hotter or colder climates.

The ancestor of the dog is correctly termed a "wolf" (or "wolf-like") because it undoubtedly descended from the same immediate ancestor as the living wolves. However, as discussed in previous Chapters, it is unlikely that the dog ancestor was the Holarctic gray wolf, the largest wolf of Eurasia, which specialized in pack hunting of very large prey. Out of all of the living species of canids only three, the gray wolf, South American bush dog and the dhole specialize in pack hunting and therefore have the stable long-term pack social structures necessary for that specialization. The Ethiopian wolf lives in community groups, but hunts small prey singly. All other canid social systems are based on a mated territorial pair that occasionally forms temporary family groups with offspring or other relatives such as siblings.[16] Despite the considerable literature that says the dog is a pack animal like the wolf, what we know of natural dog behavior indicates they form only facultative (temporary or optional) hunting groups with other individuals within a long-term "tribe" of acquaintances. Therefore, as explained in Chapter 5, until contrary evidence is available, the most parsimonious hypothesis is the dog ancestor had the common form of canid social behavior, based on a mated, territorial pair that may form temporary groups.

There is another line of support for the hypothesis that the gray wolf was not domesticated. Everywhere in the world, the generalist, mid-to-small size predator /omnivore / scavenger species are the ones that become best adapted to living around humans. Coyotes and red foxes can make a living and reproduce right in modern cities. Jackals have scavenged the refuse dumps outside of Old World villages for thousands of years. Opossums, raccoons, rats, and house mice—all omnivores—thrive in human-centered habitats. In fact, house mice (the species that was domesticated), which have co-existed with humans for about 8,000 years, now cannot survive too far from human settlements in areas where other small mammal competitors live. Like the aboriginal dogs, they have lost the instincts needed to compete in the wild.[17] Large carnivores have never become commensal for the very good reason that humans, like most potential prey, protect themselves—and most importantly their children—from the danger by threatening the carnivores to discourage them from staying near people, and killing those that persist in hanging close-by.

FIGURE 6.5. *Two aboriginal Thai dogs.*

More support for the 'not-gray-wolf ancestor' is, as discussed previously, the simple fact that large carnivores are very expensive to feed. Wolves nutritionally need considerably more meat than a human (or at least larger amounts of high quality protein). Since commensal wolves would require a lot of training before they could hunt for themselves or cooperate with humans to increase hunting success, there is no plausible scenario in which a band of Paleolithic gatherer-hunters would choose to support a population of tame wolves. On the other hand, letting a few mid-size scavengers hang around might have its advantages, which will be discussed later.

By the nature of the broad niche they live in, canid scavengers and small game predators have to be rather bold and exploratory opportunists. They have adapted to being indiscriminant in what they eat, willing to explore novel food sources, and able to thrive on often meager rations. This is why they have an upper size limit well below the big-game-hunting wolves. Strict meat eaters become easily habituated to certain prey species and find it difficult to switch to other prey. They usually will do so only when their preferred prey has fallen to such low numbers it is too "expensive" to spend energy finding it, and then they will begin to prefer the next most profitable prey. If the environment or the prey species changes too quickly ("quickly" in this case means years to tens of years), highly adapted specialists often are not able to change and are extirpated (become locally extinct). This is what happened to the largest species of *Canis* that ever lived: The North American dire wolf. When the megafauna of the Pleistocene in North America, such as mammoths, died out about 10,000 years ago this huge wolf adapted to hunting and scavenging megafauna also died out. It was replaced by the 25% smaller gray wolf, a species that first evolved in Beringia or the northern

FIGURE 6.6. *The territory of this red fox includes a park in London with trash cans to scavenge.*

Siberian tundra. The gray wolf became the top *Canis* predator of the Eurasian tundra plains and in North America after the Dire wolf extinction. Wolves have not been recorded from tropical or subtropical areas, except for the dire wolf and later the red wolf in subtropical southern North America, but did adapt to temperate forests as trees replaced grasslands after the last Ice Age.[18]

Given the above, it would be most likely that the canid that attached itself to humans was a generalist predator/scavenger, not a hyper carnivore. It must have been small enough to be considered non-threatening to humans, maybe something similar in size and behavior to the golden jackal. The golden jackal is an omnivorous mid-size canid, an opportunistic predator of birds, rodents, and other small animals that also consumes significant amounts of insects, fruit and carrion. They can serve as a representative of the general traits the original dog possibly had. These jackals are still present in North and East Africa, the Middle East, Arabia, India, SE Asia and Asia, with a few still in parts of SE Europe. At 18–20 inches (44.5–50.0 cm) in shoulder height and between 15.4–33 lbs. (7–15 kg), they are similar in size to the earliest dogs from archaeological sites, the aboriginal village dogs, small Australian dingoes, and the New Guinea dingo. This jackal can live in almost any habitat, from jungles to deserts, because it adapts its behavior to different types of prey and does not depend entirely on meat.[19]

FIGURE 6.7. *Golden jackals are extremely adaptable and opportunistic, traits the original dogs must have had. The widespread species varies in appearance over its vast range. This one is Indian.*

As described in previous Chapters, the dog has heaver bones and a much heavier skull and jaw than a jackal or coyote, indications that its ancestor had the ability to singly take somewhat large prey on a regular basis. Single jackals occasionally get lucky and are able to take down something the size of a deer or antelope fawn, but they do so with dangerous levels of stress on their jaws and teeth. With sturdier jaws and teeth, ancestral dogs could also get the marrow out of bones too large for a jackal to chew into, making the dog an even more efficient scavenger. The relatively shorter legs and broader chest of the dog compared to the wolf are indicators that it was likely an ambush predator, like the jackal, not a running predator like the wolf. The dog's large molar grinding surfaces indicate it is better adapted to eating more non-meat food than the wolf.

A mid-size generalist like the hypothetical dog ancestor would probably have been tolerated by Paleolithic humans, just as coyotes and jackals are tolerated by larger predators and modern humans. It would not have been a direct competitor for the largest game favored by humans, and the danger it presented to them was minor compared to that posed by the larger wolves: Co-existence was possible. But since even non-threatening canids tend to get into things and can be a nuisance, why did people tolerate the ancestral dogs hanging about? The critical question is why would the dog choose the new niche of scavenger around humans?

The dog's reason for associating with humans is much more complicated to answer than the human one. You have to figure out why a canid would make such a drastic change in their niche. Most animals normally do not leave their favored habitat type (e.g. woods, brush, jungle, savanna) to try an unknown habitat, if the familiar habitat is intact, unless the that habitat is saturated with competitors for the same vital resource. For example, the North American wolves that specialize on hunting caribou and follow the herds on their annual migration from the tundra to the edges of the forest and back are genetically separated from (usually do not breed with) the resident wolves in the forest that specialize on moose and elk, even though they both on occasion use the same habitat temporarily (not the exact same area that would result in territorial fighting, but the edges of the same forest). The wolves' traditional habits keep them from connecting. However, scavenger generalists like jackals—and probably ancestral wolf dogs—are specifically adapted physically and behaviorally to utilize a variety of food types and habitats. They are constantly testing new sources. Generalists will venture into the unknown, such as city streets, just to see what resources may be there. If the resources can support them, they take up residence. A wolf pack will explore beyond the usual limits of its territory as long as they are not venturing into another pack's territory. Young wolves wander far searching for places to establish a territory of their own. But usually wolves stick to familiar habitat types. Wolves will also try to take any prey they happen to encounter, but they are more creatures of tradition (previously learned behavior) than generalists. Lastly, due to their large body size and meat specialization, wolves are more limited in where they can survive than omnivores.[20]

Ancestral dogs, like the coyotes that follow wolf packs to have a steadier food supply with less effort, may at first have used humans for the same reason. This is not so unusual in wild animals. In addition to coyotes, birds such as ravens follow wolf packs on the hunt, watching to see if the wolves make a kill so they can dart in to grab what meat they can when the wolves are satiated. When Paleolithic hunters killed a large animal a long walk from their camp, they undoubtedly carried back only the most fat-rich and meatiest parts that were fairly easy to transport: Front and rear quarters, back strap with vertebrae, and maybe some organs. In leaner times, such as the cold season (or in tropical climates the dry season) when body and marrow fat are reduced in larger prey, the Paleolithic hunters probably carried the heads back too, as the brain and jaw bone contain a lot of fat. But even in the lean season, what was left at the kill site was likely more than enough to supply a mid-size scavenger with easy supplemental nutrition. If the ancestral dogs were shadowing the hunting humans, they would be among the first scavengers at the kill site, thus preempting their direct scavenger competitors. This would give the dogs a slight advantage, especially since with their stronger jaws they could easily defend the remains from jackals.[21]

The bolder ancestral dogs that could tolerate following closer to the humans would thus get the most nutritional benefit: First choice of the scraps. This relationship would be strengthened if the dogs also stayed near the camp waiting for the people to leave on a hunt. While hanging out, they could be rewarded with occasional discarded (not purposefully fed) bones, fruit rinds, or even pieces of hide. The real bonanza, though, for the camp dogs was mentioned in Chapter 3: The very nutrient rich human feces deposited around the edges of the camp. Unlike obligate carnivore wolves, the more omnivorous dogs would be attracted to the feces with its amino acids, fats and predigested carbohydrates. This perhaps partly answers the question of why Stone Age people would welcome or at least tolerate the dogs' presence. While Paleolithic people undoubtedly were not as fastidious about their personal habits as modern people, or concerned about "bad" smells, if the band consisted of the usual estimate for a strict gatherer-hunter migratory group of about 10–25 adults, over the period of several weeks in one location feces would build up to the point that insects such as flies could become annoying, especially in the warmer seasons or climates. The sanitary service the dogs provided would be a slight benefit to the people, and it only takes a small benefit for a trait or habit to be positively selected for. In this case, the behavior selected for was humans accepting co-habiting dogs.[22]

It also is probable that the people saw camp-follower dogs as a source of food when more preferred game was scarce. Why not let a potential dinner stay nearby? Surely they did not kill many of the dogs, nor make them a regular source of food as many cultures later did, or there would not have been enough camp followers to make a domesticated population. But as J. Clutton-Brock called it, having such a "walking larder" of fresh meat would have been a survival trait for the people, one which could have helped them out-compete their neighbors in lean seasons. For thousands of years before the domestication of other animals, dogs, which basically

FIGURE 6.8. *Indog dogs scavenging street side with cows.*

raise themselves on left-overs, could have served as the perfect emergency rations. Much later in the man-dog relationship, dogs came to have spiritual and religious meanings for various cultures, especially as sacrificial offerings, but it is highly doubtful these purposes had anything to do with the original reason why dogs were domesticated.[23]

A problem for this domestication without confinement or control, whether the ancestor was wolf or dog, is that the camp-follower ancestral dogs would not be genetically separated from the wild dogs. Without an effective breeding barrier between camp dogs and wild dogs, the two populations probably would regularly interbreed. If so, the wild traits would keep being reintroduced to the commensal population, preventing or greatly slowing down domestication. One possible solution to this problem is that some dogs became so habituated to the scavenger-around-humans niche that they followed the people when they moved out of the original area to places where there were no wild dogs (Paleolithic humans are believed to have been seasonally migratory). By the time the people migrated back into wild dog territory, the camp dogs would be strangers to the local wild dogs, with different habits. Like the caribou wolves and forest wolves, interbreeding would be greatly reduced. This solution rests on the proposition that the original wild dog had a restricted geographic range, which was visited seasonally by a human band.

The alternative possibility is that domestication could have taken place *in situ* within the wild dog range, even with wild dogs interbreeding with the camp follower dogs. In a rich environment like the forests and mosaic forest/grasslands of Pleistocene SE Asia, where seasons were mainly wetter and dryer rather than drastically changing, a human group with a large enough hunting territory may never have been compelled to migrate. In largely undisturbed environments, when some species are locally reduced in numbers, they are renewed by immigration from surrounding areas. This is called a "source-sink" process and the system can be maintained for extended periods, unless it is disrupted by some other factor like a wider die-off of the prey species due to disease, disaster, or prolonged drought. If the Paleo hunter-gatherers with ancestral dogs resided permanently in one area, it would have taken more time for the domestication process. But it could have proceeded like the red jungle fowl's predicted slow demise as a wild type. That is, as the human population, and therefore the proto-domestic dog population, increased slowly over thousands of years, the dogs that chose to use humans as their niche were more successful at reproduction than the wild dogs. Eventually the domestic genes overwhelmed the wild, and all the dogs thereafter (except the dingo ancestors, which were removed from the area early enough) were free-ranging domestic dogs dependent upon humans.[24]

This direction leads into the most radical speculations of my hypothesis the dog is a natural species. I think the original ancestral wolf-dog evolved in the SEA area (which includes the extreme southeast of China) where there is no evidence of other wolves, and that they made the connection to humans long before 20,000 BP. I first thought that SEA might be the area of dog origin decades ago when I started researching aboriginal

FIGURE 6.9. *An Australian dingo, one possible model for ancestral dog appearance.*

dogs. All of the village dogs in the Old World tropical and subtropical climates looked very similar. They were superficially all "dingo-like" in appearance.

Investigating Paleo human migration paths and the early ocean trade routes facilitated by currents, it seemed to me that SEA could have been a central point connecting the ancient aboriginal dog landraces of the Pacific, India, China, and Japan. Recent DNA studies of village dogs in SEA revealed that they have very high genetic diversity, and indication of their population's age, and that the DNA types found in other areas are subsets of those found in SEA. This pattern could indicate that small numbers of dogs from the SEA population founded other populations in Eurasia. SEA was also home to the most unique dog DNA types, indicating the population there has had a longer evolutionary history than populations from other areas. This supports a SEA origin for the dog, but it is a weak (secondary or indirect) support as discussed in Note 1 below.[25]

Of course, the SEA dog DNA results were first interpreted as meaning the dog originated from the gray wolves in southern China, and further that these dogs retained the original genetic diversity inherited from the wolf ancestors.[26] However, because the SEA AVDs are free-breeding with each other, not artificially divided into "breeds," the results could merely reflect the lack of human selection that narrowed genetic diversity in the comparison populations sampled. In fact, when free-ranging African AVDs were tested, they had diversity similar to the SEA AVDs, indicating that the high diversity of the SEA dogs is not unique.[27] Nevertheless, I think that the SEA area is still the likeliest candidate for where dogs originated. This area has been called a hotspot for evolution of mammals due to the number of unique endemic species.[28] The ancestral dog could have been another unique species that evolved there. One problem for the wolf origin hypothesis is that there are no gray wolf fossils reported from SE China. Although wolves have been listed as recently living there, I could find no confirmation of wolves except in more northern areas of China, nor any Pleistocene canid fossil record at all from greater SEA.

There are various reasons for the lack of archaeological dog finds in SEA before about 4,000 years BC. First, not much archaeology has been done in that area relevant to the late Pleistocene time frame of the origin of the dog. Second, fossils do not preserve well in the SEA climate (rock crevices and caves excepted). Third, much of the land available in the Pleistocene has disappeared. During a period of 10,000 years, at the lowest sea levels of the last Ice Age which peaked at about 26,000 BP, the SEA land area was expanded into a large plain with forests and rivers that ended where the Indonesian archipelago is today. Named Sundaland, or just Sunda, it connected Borneo, Sumatra and Java to Mainland SEA. Sea levels during the glacial maximum were 392 to 425 ft. (120-130 m) lower than they currently are. The lowlands of Sunda today are generally some 327 ft. (115 m) below current sea level, but at the glacial maximum would have been 65 to 98 ft. (20 to 30 m) above sea level. Around half of the land of Sunda, or an area roughly the size of India, was lost to the sea between 15,000 and 7,000 years ago when the sea level rose as glaciers retreated.[29]

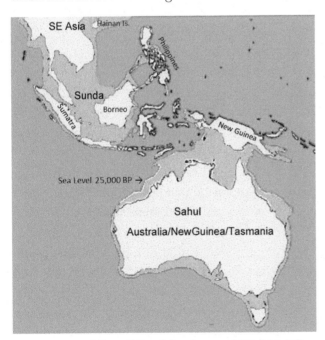

FIGURE 6.10. *Map of Sundaland around 25,000 BP. The darker tan area was land, which extended north along what is now China's east coast and encompassed Hainan island and Taiwan.*

Research on Sunda indicates the region had a mosaic of rainforest, savanna and marsh habitats, and that during the time Sunda was present mainland SEA developed corridors of grass and brush savanna. Mixes of habitats like these, which were changing significantly in relatively short periods of time (on the order of a few hundred to a few thousand years at most), would favor the evolution of a generalist canid. The small ancestral wolf I envision would have arrived in the area well before the last Ice Age, but after dholes and jackals were present. As a species, it avoided direct competition with those canids by evolving into a predator with strong jaws that could singly take prey as large as itself. The jackal was somewhat smaller than the ancestor, with weaker jaws, and the dhole hunted cooperatively in packs. The ancestral dog fitted into a unique "in-between" niche. As sea levels rose at the end of the Ice Age both the people and dogs of Sunda would have migrated to higher land, increasing the population density of both species in SEA and beyond.

EARLY DOG ANCESTOR AND HUMAN ASSOCIATION?

There is another idea about early dogs that, while not part of the natural dog hypothesis model, is interesting to contemplate. This is the highly speculative notion that the pre-domestication process for the dog could have started well before the first modern humans arrived in SEA. If one considers the dog a natural species, this idea is not outside the realm of possibility. About 50,000 BP the modern human (*Homo sapiens*) population was expanding along the coastal route from India into SEA and Australasia. But archaic humans (*Homo erectus*) had been living in the SEA area for at least a million years. The *Homo erectus* people had cultures that involved communal living, an assemblage of stone tools, control of fire, and an ability to cross large expanses of sea (a deduction based on the age of *erectus* fossils on islands and on past sea levels). Studies of *erectus* skulls show that their brains had the "Broca's area" associated with language. They may have had complicated vocalizations if not a formalized language. Male *erectus* are estimated to have averaged about 5 ft. 10 in (1.79 m) in height, with females somewhat smaller, so they were on average even taller than some modern populations.[30]

For a long time it was thought that archaic humans rapidly died out in every area after modern humans arrived because they could not compete with the more "sophisticated" moderns. But a few decades ago *erectus* fossils only about 40,000 years old were discovered in Indonesia, indicating that the two types of humans coexisted for thousands of years. Another hominid (species closely related to humans), the small extinct pre-modern people of Flores Island, Indonesia first discovered in 2004, lived there from about 80,000 –18,000 BP. Recent DNA research indicates that today's humans have a mix of DNA from least three types of human populations: *Homo sapiens*, *Homo neanderthalensis* (Neanderthal), and a distinctive population of pre-moderns (maybe a regional variety of *erectus*) from Siberia called Denisovian. The Denisovians are known only from a few bones from one site dated to about 40,000 BP, and only one sample of DNA has been extracted from a small finger bone. Even with this small comparative sample, scientists found some of the people of Melanesia (including New Guinea) share the most DNA with Denisovians, 2–4%. As research progresses and more is learned about the genetic variability of all these hominids, it may be discovered that some of what is currently thought to be modern human DNA is from other pre-modern populations.[31]

This extended presence of humans in SEA means that the division between ancestral dogs that chose the scavenger-around-humans niche and wild dogs could have started well before 35,000 years ago, the upper limit currently accepted by geneticists. This extended co-habitation with hominids could also apply to wolves in other areas, although for the reasons given in Chapter 1 this is extremely unlikely. The first estimate from DNA for the date of separation of the wolf and dog was about 125,000 BP. That date was not accepted by the scientific community, partly because modern humans had not yet spread from Africa and people have been deemed necessary for evolution of the dog from the wolf, but also because it was based on a probably incorrect assumption of one female ancestor for the relevant mtDNA clade and no hybridization with gray wolf.[32] However, an extended time-frame for the pre-domestication "following" stage makes better biological sense for both the natural dog and wolf origin hypotheses, allowing time for a slow natural evolution into domestication. Actually, if the dog did start as a gray wolf, it seems unlikely all of the physical changes from wolf to dog, especially in the skull and mandible, could have happened in the most frequently reported times for the wolf/dog genetic separation of only 15,000–20,000 years. There would have to be fairly strong natural selection for those specific characteristics, and that would happen only if these traits are adaptive for the new niche of scavenger-around-humans. Other long-term commensal scavengers such as the raccoon, opossum and Norway rat do not show such drastic changes compared to non-commensal populations. We do know some bird, fish, reptile, and other animal and insect populations have evolved adaptive traits in much shorter time-frames when they changed habitats. Niche adaptations led to one of the few undoubted cases of sympatric speciation (two populations of the same ancestral species remain in contact and yet diverge enough to be considered separate species). The multitudes of cichlid fish species in Lake Victoria, East Africa, have evolved from a single ancestral type since the lake dried out and then re-filled about 14,700 years ago, so those species that evolved there can only be that old. No one knows at this point just how long it takes for a new mammal species to evolve from an ancestor.[33]

Evolution normally acts in a very slow way, because if the change is too big, too different from the original, the animal is likely to be less functional. Its parts may not work together properly. So most of the time any physical changes are very minor. If that minor change makes the individual animal more likely to live to reproduce, that change will become more common (selected for) in the population it belongs to. For instance, once size reduction or increase has started in a population of a species due to a change in its habitat (e.g. climate) or food availability, the size will change slowly to the point where any more change makes survival harder. Then, unless the environment changes again, size will continue to vary somewhat around the body mass most adapted to that environment and diet. This evolutionary balance is why wild animals' size and shape vary so little. In the case of the dog and wolf ancestor, in the North it evolved into the large and big-toothed meat-specialist wolf, adapted physically to chasing large prey in packs. According to the Natural Species Hypothesis, in SEA the common ancestral canid stayed small and generalized, but developed a heavy jawbone, broad muzzle, and strong canine teeth as it adapted to utilizing both small and moderately large prey, and/or scavenging large bones.

If the prejudice that the dog is merely an artificial variety of wolf is suspended, then the dog can be analyzed as if it were a wild canid. Objectively compared to its nearest relative, the gray wolf, the dog is more than different enough that it would easily be declared a separate, "good" species. The dog has more differences from the wolf than there usually are between two closely related sister species (two species that evolved from the same ancestor).[34] If you take artificial selection out of the picture, the pertinent question is how long does it take on average for a population to become a new species? Unfortunately, the answer is that we have no way to know.

The rate of speciation seems to vary with the kind of analysis done and the type of animal. Rates given by paleontologists vary from 0.01 (one one-hundredth) to 10 speciation "events" (new species appearing in the fossil record) per lineage per million years. Animals with longer developmental and generation times, like elephants and humans, take longer to evolve changes compared to smaller, faster-to-mature, shorter-lived species like most canids. Of course, the appearance of a new species is stochastic, and does not happen like clockwork every so many thousands of years. But the fastest rate of 10 events per lineage per million years would mean an evolutionary line could develop a new species every 100,000 years, a time-frame considered long enough for species-level changes. J. J. Sepkoski (1998) estimated that an average of 3 new species arise per year, world-wide. This is close to the world-wide extinction level (the non-human assisted extinction level), and even if Sepkoski's estimate if off by an order of magnitude, this indicates that new species evolve rarely.[35]

Interestingly, the estimated 100,000 year interval for new species evolution is comparable to the discarded 125,000 BP split between the dog and wolf proposed by C. Vilá, et al. (1997). If the differences between dog and wolf are due only to adaptive changes from being a specialized pack hunting predator to scavenger-around-humans, without any direct artificial selection to speed things up, the time-frame would work for the wolf-origin hypothesis. The problem is, there currently is no evidence that merely becoming a scavenger would exert the selective pressure to change the wolf's skull morphology into the dog's.

As discussed previously, traits evolve because of specific reasons, due to necessity imposed by factors in the environment or a change in the animal's life-style, not because of disuse. Perhaps changing to a scavenging lifestyle would result in a smaller over-all size in camp-follower wolves as a response to reduced nutrition, but it is doubtful there would be a biological reason for changes in the shape of the lower jaw, skull, or teeth. On the other hand, there is no support for the 20,000 years of separation of the wolf and dog estimated by geneticists being long enough, even with some artificial selection to speed up changes, to result in the wolf/dog differences. Fully formed dogs are "suddenly" in the archaeological record starting about 14,000 BP. By 12,000 BP they are found in burials in many areas, an indication of how well integrated the dog had already become in human society.[36] So an origin only 20,000 years ago would mean that the much smaller dog was created from the wolf in just 6,000 years during a time when selective breeding was unknown (20,000 BP years since the supposed split, minus 14,000 years to the first agreed-upon dogs). Maybe evolution of the teeth and the skull shape are possible in that short time without direct natural selection to fit a new niche, but we have no way to know. Or, maybe the recurrent date of 20,000 BP in estimates from DNA for the separation is the result of something else, such as the dog generation time or effective population size in the formulas used in the time since separation estimates being too small, as outlined in Chapter 4.

Overall, I believe the Natural Species Hypothesis for the origin of the dog conflicts less with what is known about selection (natural and artificial), evolution, and human and canid behavior, than does the wolf-origin hypothesis. It fits with the fact that all primitive and aboriginal dogs have a similar size and appearance without specific artificial selection. It explains much more than the simple black box of the wolf origin idea, which merely says "somehow" the wolf became the dog.

NOTES

1. Definitions of hypothesis and theory. Since we cannot travel back in time and observe the beginning of the dog-human partnership, we have to rely on what scant clues we do have to make reasonable guesses about the origin of the dog. Reasonable or "educated" scientific guesses are not mere conjectures made without evidence. By the time a scientist makes a guess about something public, objective evidence must support it. To be credible, the guess has to be based on some actual objective information and cannot violate any previously identified facts. In science an idea is turned into a formal guess by writing it as a "hypothesis." A hypothesis is, according to *The Dictionary of Ecology, Evolution and Systematics* (R. Lincoln, G. Boxshall and P. Clark, eds. 1998), defined as: "An assertion or working explanation that leads to testable predictions: an assumption providing an explanation of observed facts, proposed in order to test its consequences." It is a testable possibility.

    Non-scientists often use the term "theory," as in "It's only a theory" when they mean "It's only speculation." But in science a theory means something that has lots of evidence to support it. A theory starts as a hypothesis, an idea that is stated in a specific manner. Then after the hypothesis is supported repeatedly by several kinds of evidence, it becomes a well-accepted explanation of something in Nature: A theory. Evolution is a theory. Einstein's general relativity is a theory. Nothing about origin of the dog stories rises to the level of a theory. The important thing about a good "working hypothesis" is that it can become a theory if it leads to predictions that can be subjected to testing of some sort. Good, or valid, scientific hypotheses have to be formulated so they are "falsifiable" (can be proven incorrect) by some objective evidence.

If there is no way to test the idea, it is not a scientific hypothesis. In this meaning "test" is not restricted to laboratory experiments. Test means something predicted by the hypothesis can be validated by various sources, such as information from archaeology, genetics, or paleontology. Even if a hypothesis has not yet been supported, it is still considered a "good" hypothesis if it could be tested and it has not been disproven.

What if there is more than one hypothesis that could explain something? In this case the hypothesis that has objective supporting evidence, or has the fewest assumptions to get to the same conclusion, is considered the better choice. This is the scientific principle known as Occam's razor or the "principle of parsimony." This principle states that among competing hypotheses, choose the one with the fewest assumptions.

If there are multiple hypotheses about something that have the same number of assumptions, but no available direct objective evidence, the one that has the most support from secondary sources (the "best supported" one) should be given preference. No hypothesis, or even theory, can be proven to be absolutely and for all time correct. The next discovery could disprove them. Hypotheses, and by extension theories, are either supported or falsified through an accumulation of evidence or by direct testing. When hard evidence for a hypothesis is not available and direct tests of its premise are not possible, it is perfectly acceptable to use information from indirect evidence. Indirect evidence can be, for instance, extrapolating the hunting behavior of extinct species from that of very similar living species. Close genetic similarity is indirect evidence two species are related.

2. Conroy, C. J. and M. van Tuinen, 2003; Foote, M., 2001; Muñoz-Durán, J. and B. Van Valkenburgh, 2006.

3. Brugal, J-P. and M. Boudadi-Maligne, 2011; Cherin, M., et al., 2013; Cherin, M., et al., 2014; Garrido, G. and A. Arribas, 2008; Martin, L. D., 1989; Olsen, S. J. and J. W. Olsen, 1982; Palmqvist, P., A. Arribas and B. Martinez-Navarro, 1999; Pei, W-C., 1934; Sardella, R. and M. R. Palombo, 2007; Sardella, R., et al., 2013; Sommer, R. and N. Benecke, 2005; Sotnikova, M., 2001; Tong, H. W., N. Hu and X. Wang, 2012.

4. Anderson, A. J., 1990; Burleigh, R., et al., 1977; Clark, K. M., 2000; Davis, S. and F. R. Valla, 1978; Dayan, T., 1994; Dayan, T. and E. Galili, 2001; Gollan, K, 1982; Gonzalez, T., 2012; Haag, W. G., 1948; Harcourt, R. A., 1974; Higham, C. F.W., et al., 1980; Horard-Herbin, M-P., A. Tresset and J-D. Vigne, 2014; Lawrence, B., 1967; Morey, D. F., 1986; Musil, Rudolf, 2001; Napierala, H. and H-P. Uerpmann, 2010; Olsen, S.J., J.W. Olsen and Q. Guo-qin, 1980; Pionnier-Capitan, M. et al., 2011; Shigehara, N., 1994; Shigehara, N., et al., 1993;

Shigehara, N., et al., 1998; Shigehara, N. and H. Hongo, 2001; Tchernov, E. and F. F. Valla, 1997.

5. Van Valkenburgh, B. and K. P. Koepfli, 1993.

6. Coyne, J. A. and H. A. Orr, 2004; Crandall, K. A., et al., 2000; Doebeli M., and U. Dieckmann, 2002; Mayr, E. & Ashlock, P. D., 1991; Muñoz-Fuentes, V., et al., 2009; Thompson, J. N., 1998; Via, S., 2001; Via, S., 2012.

7. Reponen, S. E., et al., 2014.

8. Gittleman, J. L., 2013; Moehlman, P. D., 1983.

9. Sacks, B., et al., 2013; Savolainen, P., et al., 2004; Oskarsson, M. C. R., et al., 2011. The time estimates from DNA are used in this book with the acknowledgement of the potential problems of the methods discussed in Chapter 4. Because all of the samples of dog and wolf used in an individual study are analyzed in exactly the same way, even if the actual time span estimate is off, the relative ages of genetic lines within that study would be informative.

10. https://en.wikipedia.org/wiki/Dingo_Fence

11. Paquet, P. C., 1992.

12. Evin, A., et al., 2013; Larson, G., et al., 2012; Lau, A. N. et al., 2008; Reed, C. A., 1954; Zeder, M., 2012.

13. Liu, Y. P., et al., 2006; Sawai, H., et al., 2010.

14. Majumder, A., et al., 2011. I listed only one reference, but I read several recent reports of mammal surveys in SEA and eastern India, mostly camera trap studies, and corresponded with field researchers. The few dogs reported deep in the reserves were listed under "human disturbance" because they only traveled in the interior of reserves with hunters there illegally. Otherwise dogs, like those in Africa, only go short distances into reserves.

15. Corbett, L. K., 2004.

16. Sillero-Zubiri, C., M. Hoffmann and D. Macdonald, eds., 2004.

17. Connor, J. L., 1975.

18. Anyonge, W., and Baker, A., 2006; Fuller, T. K., et al., 1989; García, N. and E. Vírgós, 2007; Geffen E., et al., 1996; Gittleman, J. L., 1985; Gittleman, J. L., 2013; Holliday, J. A. and S. J. Steppan, 2004; Muñoz-Fuentes, V., et al., 2009; Nowak, R. M., 1979; Sacks, B. N., et al., 2008; Van Valkenburgh, B. and R. K. Wayne, 1994.

19. Gittleman, J. L., 2013; Macdonald, D. W., 1979; Majumder, A., et al., 2011; Poché, R. M., et al., 1987.

20. Hallett, M., 1987; Musiani, M., et al., 2007; Pilot, M. et al., 2006; Saunders, G., et al., 1993; Sillero-Zubiri, C., M. Hoffmann and D. Macdonald, Eds., 2004;

Weckworth, B. V., et al., 2011.

21. Churchill, S. E., 1993; Gifford-Gonzalez, D., 1991; Grayson, D. K., 2014; Lee, R. B. and R. H. Daly, Eds., 1999; McDonald Pavelka, M. S., 2007; Stiner, M. C., 1994; Stiner, M. C., N. D. Munro, and T. A. Surovell, 2000; Straus, L.G., et al., eds., 1996.

22. Butler, J.R.A. and J.T. du Toit, 2002; Butzer, K. W., 1982; Lee, R. B. and R. H. Daly, eds., 1999; Speth, J. D. and K. A. Spielmann, 1983; Wolpoff, M. H., 1999;

23. Clutton-Brock, J., ed., 2014; Martín, P., 2014; Kolig, E., 1978; Leach, M., 1961; Podberscek, A. L., 2009; Schwartz, M., 1998. Many of the references in Note 4 for prehistoric dogs and wolves report indications the dogs were eaten.

24. Amarasekare, P. and R. M. Nisbet, 2001; Cannon, C. H., R. J. Morely and A. B. G. Bush, 2009; Clark, P. U., 2012; Heaney, L. R., 1986; Louys, J. and E. Meijaard, 2010; Meijaard, E., 2003; Slik, J. W. F., et al., 2011; Wurster, C. M., et al., 2010.

25. Forster, P. and S. Matsumura, 2005; Kahlke, R-D., et al., 2011; O'Connor, S., R. Onon and C. Clarkson, 2011; Soares, P., et al., 2008.

26. Ding, Z-L., et al., 2012.

27. Boyko, A. R., et al., 2009.

28. Zachos, F. E. and J. C. Habel, eds., 2011; http://www.cepf.net/resources/hotspots/Asia-Pacific/Pages/Sundaland.aspx

29. sundalandhttps://www.eeb.ucla.edu/Faculty/Barber/Animations.htm; Cannon, C. H., R. J. Morely and A. B. G. Bush, 2009; Cranbrook, E., 1988; Hanebuth, T. J. J. and K. Stattegger, 2010; Houghton, P., 1996; James, H. V. A. and M. D. Petraglia, 2005; Morrison, K. D. and L. L. Junker, 2002; Sathiamurthy, E. and H. K. Voris, 2006.

30. Forster, P. and S. Matsumura, 2005; Jinam, T. A., et al., 2012; Joordens J.C.A., et al., 2015; Marwick, B., 2009; van den Bergh, G. D., 1999.

31. Finlayson, C., 2009; Hawks, J., 2013; Morwood, M. J., et al., 2005; Vinayak, E., H. Harpending and A.R. Rogers, 2005.

32. Vilà, C., et al., 1997.

33. Grant, P. R. and B. R. Grant, 2002; Meloro, C. and P. Raia, 2010; Samonte, I. E., et al., 2007; Verheyen, E., et al., 2003; and references in Note 36.

34. Coyne, J. A. and H. A. Orr, 2004; Mayr, E. & Ashlock, P. D., 1991.

35. Avise, J. C., D. Walker and G. C. Johns, 1998; Brugal, J-P. and M. Boudadi-Maligne, 2011; Larkin, J. C., 2009; Sepkoski, J. J., Jr., 1998.

36. Cole, B. F. and P. E. Koerper, 2002; Morey, D. F., 2010; Morey, D. F. and Michael D. Wiant, 1992.

CHAPTER 7. CONCLUSION

What is wanted is not the will to believe, but the wish to find out, which is the exact opposite. —Bertrand Russell. in *Sceptical Essays* (1928).

Here again from the introduction are the currently irrefutable, scientifically proven facts about the origin of the dog:
1. The dog and gray wolf are very closely related.
2. The oldest recognized dog fossils are dated to about 14,000 years ago.

All the speculation about the gray wolf, *Canis lupus*, being the ancestor of the dog, *Canis familiaris*, is based solely on fact 1. However, the gray wolf ancestor hypothesis is contradicted by the danger posed by wolves that associate humans with food, a strong reason prehistoric humans would not have allowed them to hang about camps, and the lack of any viable beneficial reason for humans to co-habit with gray wolves. If the differences in the dog and gray wolf skulls, until now explained away without serious research as domestication effects, are examined as possibly ancestral they indicate the dog ancestor was a wolf-like canid adapted to a specialized niche.

Although at present there are no identified dog remains from before about 14,000 years ago, the Natural Species Hypothesis for the origin of the dog is supported by many types of secondary evidence from natural dogs (dingoes and aboriginal village dogs–hereafter the only populations included in "dog").

1. The dog skull, jaw and teeth indicate the ancestor was a canid adapted to singly and habitually taking prey greater than 25% of its own weight, possibly using a holding bite. This niche is not filled by any living *Canis* species.

2. Dog body size is slightly larger than the small prey (less than 25% of canid body weight) adapted generalist coyote and jackals, but never exceeds a body mass that can be supported on small prey.

3. Dogs have annual breeding cycles adapted to environmental cues, like all wild *Canis* species. Because their estrus cycles are at different times of year than those of *Canis lupus*, primarily during the northern hemisphere summer, they could not have inherited that cycle from the wolf.

4. Mature dogs have fully developed canid behavior, not the juvenile, truncated, or disrupted behavior often attributed to some modern domestic breed dogs. Village dogs are perfectly adapted to surviving and reproducing on their own by scavenging around humans, and the least (or not at all) domesticated dogs, the dingoes, are wild predators.

5. From the little natural social behavior data available, dogs seem to typically be single foragers and facultative pack formers, like all Canis species except

FIGURE 7.1. *A free-ranging Indog nursing pups.*

the largest, Canis lupus. This is the most generalized and adaptable form of social canid behavior, and therefore likely more representative of the common ancestor than that of the highly specialized gray wolf.

In the past, many attributed the differences between dog and wolf to paedomorphism or neoteny, to side effects of selection for tameness (the supposed mechanism of domestication), or to mere disuse by the domestic dog. However, there is no proof that typical dog traits are the result of anything but natural selection. The current common argument consists of:

Gray wolf + Domestication + [Black Box Magic] = Dog

The Natural Species Hypothesis fills in the black box with the biologically based processes of evolution, ecology and adaptation.

Some of the purported "domestication effects," like the smaller relative brain size of the dog compared to Holarctic gray wolf, are normal traits for smaller canids the size of natural dogs. Other domestication effect claims, such as the smaller relative canine tooth size in the dog compared to the wolf, are assumptions that when examined proved to be incorrect. Because nearly all dog morphology and behavior research to date has been based on modern domestic breed or mixed-breed dogs, the conclusions they have been altered physically and behaviorally from a more natural pre-domestication ancestor are undoubtedly valid. How could it be otherwise when their ancestors were subjected to sometimes severe artificial selection for traits humans desired, often traits

FIGURE 7.2. *Like its ancestors have done for thousands of years, this Thai aboriginal dog waits nearby as a man, probably his owner, repairs his net.*

that would be non-adaptive in the wild? For the domestication effect conclusions to apply to *Canis familiaris* in general, as the authors imply, the more natural dogs, the dingoes and aboriginal dogs, need to be included in the studies. It would be most interesting to know how the natural dogs compare to the modern breed and mix dogs on the various tests and evaluations. At minimum, the comparison might tell us what behaviors have been altered by artificial selection over the last 5,000 years. That would be a good starting point for revealing basic traits that humans changed to make the dog a companion and help mate rather than just the village scavenger.

One of the most basic rules of scientific research is to ask the right question for the answer you are seeking. The research question the domestication researchers asked was "How does the dog differ from the wolf?" The answers they found were valid for that comparison. The problem is, the question actually intended was "How does the dog today differ from its ancestor?" That was the question they thought they were answering based on the unproven assumption that the wolf was the direct ancestor of the dog.

Another basic rule of scientific research is to choose appropriate subjects to answer the question. The answers about domestication effects can only be determined in comparison to an ancestor, either one proven directly or a virtual model of the most likely characteristics of the ancestor as extrapolated from secondary evidence. The only dogs capable of providing insight into the original dog (whatever origin hypothesis is proposed) are those who have been subjected to the least artificial selection, the dingoes and ancient races of aboriginal village dogs. The few studies done so far on these natural dogs indicate their behavior and morphology are more like similar-sized, less closely related species than they are like the gray wolf. The number of different populations of aboriginal dogs systematically observed so far in their natural habitats is tiny compared to the huge geographic range of aboriginal dogs. Many more studies of the behavior of free-ranging, free-breeding village dogs are needed before there can be a strong inference about behaviors that could be ancestral, other than those common to all canids. The social behavior of wild dingoes, the most likely populations to retain ancestral traits, has not been studied at all.

The genetic analyses that report the wolf and dog separated only 32,000 years ago or less, short time-frames which supposedly support the gray wolf origin hypothesis, are problematic due to the many estimates in the formulas. Slight changes in the estimations of important factors, such as the assumed mutation rate and the founding population size for dogs, have large effects on dating the separation time. At this point, such dates are questionable at best, and at worst misleading. According to mtDNA and some partial analyses of portions of nDNA, the dog and gray wolf are closest relatives, much closer than most other related species. This does not mean the dog descended from that wolf, just that they had a recent common ancestor. As geneticists are able to validate their formulas using real-life organisms in laboratory or field studies, their techniques will improve and results will become more precise. Until then, DNA dating should be considered questionable and not evidence the dog descended directly from the gray wolf.

FIGURE 7.3. *This young male New Guinea dingo from the captive population is a fine example of a natural dog adapted to the predator niche.*

The lack of recognized (identified) fossil dogs from before about 14,000 BP has been used to argue against the dog being a naturally evolved species. However, many living species have no fossil record. In the past there were numerous now extinct small varieties of wolf in Eurasia. Most probably have no direct descendants. One was the ancestor of the much larger gray wolf. Another descendant of that same ancestor could be the dog, which is about the same size as some of the extinct wolves. Today the only small "true" wolves (as differentiated from the coyote and jackals) that remain are the newly re-discovered African wolf, *Canis lupus lupaster* and, according to the Natural Species Hypothesis, the dog.

Science is a human endeavor, conducted by people who are subject to unrecognized biased thinking. When looking at evidence related to the hypothesis they are working with researchers have to guard against unconsciously ignoring data that does not fit the hypothesis being investigated, and actively investigate questions that might yield contradictory evidence. Because I am an independent scholar not associated with an academic institution, I knew my controversial Natural Species Hypothesis would be subjected to more than usual scrutiny. I did not want to inadvertently miss something that would falsify the Natural Species Hypothesis and kept both the wolf origin and separate species ideas in mind, accepting the fact that the results of my inquiries might support either one. Over the two decades I investigated questions about the origin of the dog, I sought out research that might directly support the domestication of the gray wolf. I could find none. Because I felt they would provide better information about the dog ancestor than modern breed dogs, I promoted sampling dingoes and aboriginal dogs for genetic studies asking questions about the dog's origin. Once including these dogs was proposed, the geneticists immediately saw the possibilities and found ways to fund gathering samples from aboriginal dogs. The natural dogs proved to be generally the same genetically as modern dogs, very close to the gray wolf but identifiable as dog.

After I realized no one had actually compared the dog cranial and dental forms to those of other canids objectively, I spent months of spare time measuring dog skulls and teeth (and coyote, gray wolf and red wolf for comparison), not knowing what the result would be. I was pleasantly surprised when I did the final calculations and the conclusion was that the dog is different. It did not match any published set of adaptive traits for ecological categories previously defined for canids. In fact, the dog, when looked at as a natural species and not an artificial construct, seems to be unique among living canids.

More recently, I became curious about canid howls. I wondered if howls are species specific and could therefore be used as another characteristic to define species and subspecies. Although I thought I heard significant differences between modern dog, wolf and New Guinea dingo howls, what seems very different subjectively may with systematic analysis prove to have only insignificant differences. While I found several references with qualitative descriptions of some canid howls, I could find no published research that analyzed howl frequencies and forms to compare species. My curiosity about this question motivated me over several years to keep searching for someone qualified in acoustics and mathematics who would compare the New Guinea dingo to wolves and domestic dogs. Arik Kershenbaum accepted the challenge of a comparative study starting in 2014, then took it several steps farther. First he set up a citizen science web site for diagramming howl spectrograms. Then he organized an international group of researchers interested in carnivore vocalizations, expanding the scope to include any species of cooperative carnivore. The groups' preliminary results on canid howls indicate some modern dogs and the New Guinea dingo have "similar" but detectably different howl types, and these are distinct in some traits compared to those of the three wolf species included so far.[1] Once again, when compared objectively without preconceptions, the dog and dingoes have diagnostic characteristics that are unlikely to be the result of direct artificial selection.

Everything I found in the scientific literature, the research I did myself, and the research I encouraged, have all supported,

FIGURE 7.4. *A female Australian dingo sings her song.*

or at least did not contradict, the idea the dog is a natural species. Perhaps the next discovery in paleontology or genetics will disprove the Natural Species Hypothesis. Or, with newly discovered information, perhaps someone will develop a superior hypothesis that fits with all known facts. Until that time, the idea that the dog was a dog when it attached itself to human communities agrees with all available data and biological principles. It is based on fewer assumptions and has more secondary support than the gray wolf origin idea.

Although a minority has always questioned the gray wolf ancestry of the dog, few have expressed their lack of faith publically.[2] I am confident this will soon change as more scientists realize there are other viable possibilities. If the gray wolf origin bias can be overcome, additional research and re-interpretation of previous research will follow. As presented in this book, the Natural Species Hypothesis utilizes only accepted biological and ecological concepts and peer-reviewed information, except for the speculation that South East Asia (including Sundaland) could be the location where dogs evolved. By making explicit both the unfounded assumptions—the questionable ideas—behind the gray wolf origin hypothesis, and providing realistic alternative ideas, the possibilities in this book can serve as a starting point for further discussion and constructive debate.

Very soon we will know much more about the dog's relationships to other canids. There is an international consortium of geneticists, anthropologists, paleontologists, and biologists now doing the most complete study to date of both living and prehistoric dogs and wolves. The group, organized by Greger Larson, should have first results published in 2016.[3] Also, a recent canid genetic study based on the largest sample yet of various wolf subspecies has concluded the dog did not descend directly from *Canis lupus*, but from a closely related "extinct" wolf. The dog origin location is still open to interpretation. The latest genetic studies have claimed both southern East Asian and central Asian (Nepal, Mongolia) origins based on different portions of DNA.[4] G.-D. Wang, et al. (2015) date the dog/wolf separation at about 33,000 BP. They state (p. 8):

> One possible explanation for the 33,000-year deep divergence between dogs and wolves is that it represents a split among wolf populations, and that South Chinese wolves (ancestors to the dog) were genetically differentiated from the more northern wolves sampled in our study. It is possible that the ecological niche unique in southern East Asia provided an optimal refuge for both humans and the ancestors of dogs during the last glacial period (110–12k years ago, with a peak between 26, 500 and 19, 000 years ago).

What, exactly, Wang et al. are referring to as the "South Chinese wolf" is unclear as there is no fossil record I could find of gray wolves in SE. China, a tropical area. Perhaps the only wolf there was the common ancestor of gray wolf and dog. If so, their conclusion sounds a lot like the Natural Species Hypothesis.

The obvious implication of the Natural Species Hypothesis is that the ancestral wolf is *not* truly extinct. It lives on in various aspects of the dingoes and ancient races of free-ranging, free-breeding village dog. Unless these dogs are morphologically different enough from the ancestor to be recognized as another species, the ancestor is far from gone. Available evidence indicates the generalized dog skull and teeth have not changed significantly during 14,000 years of association with man, and in the case of the dingoes, there are no significant differences from village dogs after at least 5,000 years as wild predators. The persistence of the diagnostic traits in all naturally selected village dog and dingo populations under very different environmental conditions and niches is a strong indication they are ancestral. The descendants of the ancestor, of that small wolf-like canid that chose to switch its niche and throw its fate in with humans, have become the most successful canid in the history of the world. *Canis familiaris* has, except for the two wild dingo subspecies, adapted completely to the scavenger-around-humans niche. Today, even puppies taken from the free-breeding aboriginal populations before their sensitive socialization period ends can adapt behaviorally to become co-workers, help-mates and household pets integrated into human families. The human-dog partnership will undoubtedly continue for as long as both our species last. Genetics may someday pin down the geographic origin of the dog, but only researching the behavior of aboriginal dogs and dingoes will give us a clue about why the original dog was able to meld so completely into human society.

FIGURE 7.5. *The strong bond between Kandu's Krusier, a male Rhodesian Ridgeback bred by the author, and his young human friend, Andrew Bennett, shows why the dog human partnership is unlikely to end.*

NOTES

1. A. Kershenbaum, The Cooperative Carnivore Vocalization Project, personal communication, 2015; Kershenbaum, A., et al., 2016. The howl project site where anyone can participate in the research is: http://howlcorder.appspot.com/about.html

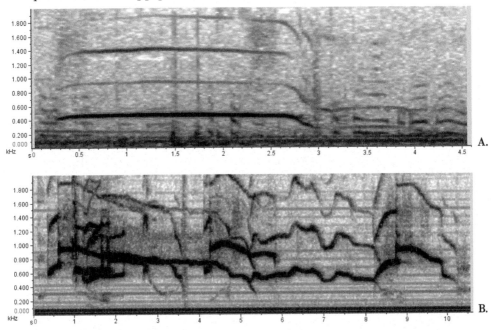

FIGURE 7.6. *Samples of sonograms of howls of (A) McKenzie Valley gray wolves and (B) New Guinea dingoes.* Courtesy of Arik Kershenbaum.

2. Fox, M. W., 1973, 1978; Dinets, V., 2015; Koler-Matznick, J., 2002; Manwell, C. and C. M. A. Baker, 1983.

3. http://dogarchaeology.org/

4. Freedman, A. H., et al., 2014; Grimm, D., 2015; Pilot, M., et al., 2015; Shannon, M., et al., 2015; Wang, D.-G., et al., 2015.

Adolescent Australian Dingo

PART II

Examples of Primitive and Aboriginal Dogs

INTRODUCTION

THE DINGOES

The New Guinea and Australian dingoes are the only two dog populations today that can be considered truly wild. Each subspecies is the top predator in those two land ecosystems and is a self-perpetuating population (they reproduce and survive in the wild). If the dog is considered a valid species, after thousands of years of geographic isolation from other dogs during which the dingoes evolved unique traits, they meet the definition of a biological subspecies. All other dogs are domesticated, and even if free-ranging, survive mainly as scavengers around humans. Because the dingoes have at least 4,000 years less association with people than modern domesticated dogs, they are important remnants of the original dog

THE ABORIGINAL DOGS

Aboriginal dogs, as defined in this book, are:
- A class of domestic dogs that emerged as an ecotype within a specific ecological niche;
- Largely the result of environmental adaptation, mostly under conditions of natural selection but influenced by human preferences and interference;
- Fit the requirements of a specific human society living in a particular ecosystem.

Johan and Edith Gallant[1]

Aboriginal dogs:
- Have evolved by natural selection under conditions of free life and close interactions with people;
- are a unique piece of nature, time bound and place bound, most similar to zoological subspecies;
- are historically associated with ethnic groups and cultures;
- are one of the oldest and most natural dogs in existence.

Vladimir Beregovoy, PhD[2]

In many places in Africa, the Middle East, Arabia, India, South East Asia (including far south east China), and the Pacific Island nations, there are ancient races of dog that have, since dogs and humans first connected, led their own independent lives side-by-

FIGURE 1.1. *A captive Australian (rear) and New Guinea dingo illustrate the typical size difference of these two subspecies.*

side with people. While some are owned, and a small minority serve humans as flock guards, household warning systems, or hunting aids and therefore are purposefully fed, as explained in Chapter 5 most make their own living. They keep the villages cleaned up of garbage and feces, a valuable service where there is no waste collection or sanitary facilities. Their survival is just as tough as if they were wild, maybe more so as they deal with human predators (surplus dog killing in developing countries with few resources, often by gruesome means such as poisoning or clubbing) and motorized vehicles, in addition to poisonous snakes, weather extremes, food fluctuations, and predators in the areas just outside the villages. For example, the diet of 21 leopards that live in Sanjay Gandhi National Park adjacent to Mumbai, India, is about 60% dog.[3]

These aboriginal dogs are not "juvenilized wolves" but intelligent, problem-solving, perfectly adapted commensal canids. They thrive and reproduce on the most meager, protein deficient diets. Their bodies are resistant to the local parasites and only become overwhelmed by mange or worms when stressed by other factors such as poor nutrition. Natural selection quickly eliminates individuals with orthopedic and metabolic problems. If brought into homes as puppies they grow into loyal, generally long-lived companions. In other words, except for some highly specialized types of work, they are actually the ideal dogs.

Like dingoes, aboriginal dogs are not, however, the ideal pet for most people in developed, industrialized areas precisely because they have extremely high intelligence and independent temperaments. Today most people are accustomed to dogs that have been specifically selected to be companions or at-command working dogs. These modern breed and mixed breed dogs are in general much easier to keep confined and less emotionally reactive (sensitive) than aboriginal dogs. Even so, there has always been a minority who are willing to adjust their lives to accommodate aboriginal dog companions, precisely because they appreciate the intelligence and independent attitude of these ancient races. For these people, other dogs, while interesting and admired, are "too domesticated." Modern derived dogs are like children who depend on us for their lives. Aboriginal dogs are like friends with whom we have a mutual affection and devotion.

There are some kennel club registered breeds that were developed from aboriginal landraces. In general they still retain the native intelligence and independent attitude of the village dogs although for some breeds the reactive temperament has been toned down, and their physical appearance changed from the free-ranging population to varying degrees (what the breeders term "improving" the breed, by selecting for instance for longer hair, greater or lesser body size, specific colors). Usually these breeds were started by a very small sample of the original landrace chosen by non-natives according to their personal aesthetic preferences and then taken out of their natural environments as founders for "pure breeds." Some examples are: the Basenji, Samoyed, Siberian husky, Finnish Spitz, Russo-European Laika, Karelian Bear Dog, the traditional gazehounds selected from different landrace breeds (e.g. Afghan hounds,

FIGURE 1.2. *This is an area outside an Indian city constructed for the natural disposal of animal carcasses. The free-ranging dogs, which appear to be Indog mixes, clean the bones, which are sold to become agricultural bone meal.*

FIGURE 1.3. *This is an aboriginal Thai dog mating with a mixed breed dog. Both the pure Australian dingo and aboriginal landraces are slowly being genetically compromised by imported modern dogs.*

sloughi, saluqui, Azawakh, Indian caravan hounds, Portuguese podencos), the German spitz breeds, Norrbottenspets, Swedish vallhund, Norwegian buhund, and Mexican xoloitzcuintli/Chinese crested.

Due to the unavoidable inbreeding in closed gene pools (no specimens from non-kennel club registered parents can be admitted to most stud books after the founders are entered), loss of genetic diversity due to selection of only a few individuals of each generation to breed, and the unfortunate human tendency to select for exaggerated traits (the proverbial "if a little is good, more is better" syndrome) before health and longevity, many pure breeds are experiencing high rates of health and reproductive problems and unnaturally short life spans.[4] Because they recently originated from a population under natural selection, most of the breeds derived from aboriginal landraces have excellent health, with a few exceptions due to disease caused by mutated genes that are concentrated in closed gene pools. If a breed gets into a genetic dead end hopefully the fanciers will insist on opening the stud book to incorporate new landrace specimens, as they did for the Basenji, do for the Canaan dog, and are in process of doing for the saluqui.[5]

The free-ranging, free-breeding aboriginal dogs of the world are a reservoir of genetic diversity for the species *Canis familiaris*, waiting to be tapped as needed.[6] At least they are until the host cultures change and disallow free-ranging dogs, or they are genetically swamped by cross breeding with imported modern domestic dogs. Today, most free-ranging dogs within several miles of any city in areas that traditionally had aboriginal dogs are mixed. But in more rural areas there are still pockets of pure ancient line dogs. Some of them are pictured in the following accounts written by people working to conserve them. The album illustrates additional landraces and breeds created from landraces. With current rates of human population growth, accelerating expansion of developing economies, and improved transportation, it is only a matter of time before these precious remnants of original dogs are lost. And that time is short.

The individual entries for the dingoes and aboriginal dogs were composed by experts. Within a basic format and word limit, they chose what to include.

NOTES

1. Gallant, J., 2002.
2. Beregovoy, V., 2001.
3. Edgaonkar, A. and R. Chellam, 2002.
4. Calboli, F. C., et al.; Leroy, G., 2011.
5. https://www.basenji.org/african/project.htm; http://www.desertbred.org/
6. Boyko, A. R., et al., 2009; Savolainen, P., et al., 2002.

THE NEW GUINEA DINGO

(Canis familiaris hallstromi)

BY JANICE KOLER-MATZNICK

FIGURE 1. *This photo was taken at the Tari Gap pass, Southern Highlands, Papua New Guinea in about 1995. The photographer asked the boy, who was passing by with his family, if he could take his picture, but failed to ask what tribe he belonged to. This appears to be a pure New Guinea dingo and may have been caught in the wild as a puppy.*

New Guinea, located just north of Australia, is the second largest island in the world. It has rugged mountain ranges tall enough to have permanent glaciers. The first non-natives, who landed in a plane in the central valley, made it to the interior only in 1910. The first true expeditions started about 1938. Many remote areas have never been explored by non-natives. The mountains are home to the New Guinea dingo (NGD), also called the New Guinea singing dog, a named given for its chorus howl. The NGD chorus howl is different from other canids' communal howls because they are not merely all howling at the same time, but appear to be actually coordinating their howls, counterpointing each other, and creating what are known in human singing as "overtones" or "ringing chords" in which voices combine creating an added note.

Like the Australian dingo, the NGD is an ancient evolutionary line of early dog and is the top land predator in New Guinea. After its introduction the NGD has been evolving under only natural selection and can be considered a true subspecies, not a "breed." Almost all we know about NGDs has come from the study and observation of captives. There are perhaps

FIGURE 2. *The Mt. Giluwe peaks in Papua New Guinea, one of the locations still reporting wild New Guinea dingoes.*

300 captive NGDs today, all descended from a few brought out of Papua New Guinea (PNG) in 1955 and West Papua in the 1970s. Almost all of the captives are in North America, although they used to be kept in many Australian and European zoos, and at the London Zoo.

Extremely little is known about the wild population, except that there are still wild NGDs in some remote places. We do know that the NGD is the top land predator of New Guinea and that its prey includes cuscus (an opossum-type marsupial), echidnas, and birds. They use natural rock and root formations as resting dens and may be seasonally migratory within their territory. Always reported to be extremely shy of humans, they are difficult to observe for more than a few seconds as they run away.

GEOGRAPHIC RANGE

NGDs have been sighted from 6,000 to 11,000 feet (2,000–4,700 m) altitude in the New Guinea mountains. Their range includes cloud forest, sub-alpine, and alpine habitats.

PHYSICAL DESCRIPTION

Captive NGDs are from 16 – 19 in. (40 – 48 cm) tall at the shoulder and 20 – 30 lbs. (9 – 13 kg). The coat is double, with seasonally thick underfur and a coarse dirt and water shedding outer coat, which is longer on the neck and the rear of the hind legs. The

FIGURE 3. *The original pair (female barely visible on the left) of NGDs brought out of Papua New Guinea in 1955, on the boat going to Australia where they were displayed at the Taronga Zoo.*

tail is carried in an upward curve and has a long fluffy brush on the underside that is a lighter color (cream or pale tan) than the rest of the coat. Two colors are known from the few sightings, the two photos of wild NGDs available, and the captive population: red sable (golden tan to fox red with some dark banded hairs on the back, top of the tail, back of the ears) or black and tan (main body is black with tan on the sides of the face, the legs and the underside of the tail). Almost all NGDs have a white tail tip and a spot or stripe of white under the chin. Many also have "Irish" white markings on the face, neck, feet, and legs. The wild population may have a broader range of color variation but their size is probably within the captive range.

BREEDING SEASON

Captive NGDs have one annual breeding season starting in August. Although reports of wild pups are rare, the approximate ages of those reported corresponds to this seasonal cycle. One of the most unique characteristics of NGDs, determined from many years of records on multiple females and by hormone assays, is that 30% of the females that do not get pregnant on the first cycle come in again into a full estrus about three months after the first ends. Then, about 10% of those females with second cycles who do not get pregnant come in for a third. No other canid has repeated estrus periods within one annual breeding season.

FIGURE 4. *The only photo to date of a live wild NGD. This black and tan specimen was seen by mammologist Tim Flannery during an expedition to Dokfuma in the Star Mountains in the 1980s. The photo first appeared in his book The Mammals of New Guinea. Because the original was lost, this was scanned from that book.*

CULTURAL ASSOCIATION

The NGD has, as far as is known, never been kept merely as a "pet" by the indigenous people. The wild NGDs reportedly avoid human habitation and never voluntarily become scavengers around villages. Before domesticated dogs became readily available, wild pups were captured and raised by a specific male hunter in the men's longhouse to become hunting aids. Reportedly, only one or two males were kept in each village group. While some lowland tribes that have been influenced by Austronesian cultural habits eat dog, most of the New Guinea Highlanders do not.

There is a rich cultural mythology about the wild NGDs, which locals clearly distinguish

FIGURE 5. *A 12 week old captive male New Guinea dingo puppy. NGDs grow very rapidly their first four months and lose the loose skin.*

from domesticated dogs. Some "elder stories" relate how the wild dog brought fire, or language, and even that the people originated through the mating of wild dog males with human females. These same myth themes are present in many aboriginal cultures around the world. This may be an indication the basic stories were developed before modern humans spread around the globe. How else can one explain the 'dog brought fire' myth being present in Africa and Papua New Guinea? One interesting Papuan myth is that the spirits of dead people inhabit the wild dogs, which are therefore considered sacred and taboo.

THE FUTURE

The first ever field studies of wild NGDs are being planned. Two sites, Mt. Wilhelm, the tallest mountain in Papua New Guinea, and Mt. Giluwe (pictured above in Figure 2), the second tallest mountain, will be surveyed for the wild NGDs. The goal is to learn basic information about the wild population and to determine how NGD predation may be affecting the prey species, especially the endangered echidnas and rare species of cuscus.

For more information: http://newguinea-singing-dog-conservation.org/

THE DINGO OF AUSTRALIA

Canis dingo

BY LYN WATSON

FIGURE 1. *A desert ecomorph dingo.*

The Australian dingo does not fit comfortably into either the domesticated dog or wolf categories. It remains an enigma. Perhaps this dog represents the nearest living relic of the Palaeolithic dog ancestor. The Australian dingo (AD) has complete canid survival instincts and is capable of a totally wild and independent life remote from any human influence, even when captive raised.

ORIGIN

The common assumption is that the dingo arrived 4,000–5,000 years ago via sea traders. The few dingo fossil remains uncovered so far support this minimum timeline, but there is no evidence for that mode of arrival. The aboriginal people, who have no written history other than cave art, disagree with the sea-farer theory, claiming the dingo has always been with them. Recently some aboriginal rock art depicting dingoes has been carbon dated to 26,000–30,000 years BP (Before Present), supporting the possibility

FIGURE 2. *A captive black and tan alpine ecomorph dingo.*

that the AD came from the South East Asian continent of Sahul *(see Chapter 6)* during the end of the last Ice age, before Australia and New Guinea separated. Perhaps the dingoes came overland with the first people.

GENERAL DESCRIPTION

The AD resembles the aboriginal village dog landraces. However, the AD has traits which set it apart from domesticated dogs. The AD does not bark like modern domestic dogs, but howls, chortles, yodels, and purrs. The AD's prey drive is irrepressible; they are consummate predators. ADs breed annually, the seasonal timing varying somewhat over the vast continent, but usually between March and May. ADs have no body odour.

The AD's obliquely angled eyes detect movement within a 180 degree view. Its hearing and sense of smell are hyper sensitive. Nothing goes unnoticed. In addition, it has extra-long vibrissae (whiskers) for tactile sensory detection. The AD's ability to flex all joints like a contortionist and generally to act very cat-like, is remarkable.

PHYSICAL DESCRIPTION

The AD is medium sized, 19-22in. (48-56 cm) in shoulder height and at maturity weighs approximately 30-45 lbs. (17-20 kg). AD coats may be solid cream, pale apricot, sandy, ginger, or bicoloured – black and tan. All colours may have some small white markings at extremities and chest. Eyes are always brown. Creams may have

FIGURE 3. *A lovely young ginger color phase alpine ecomorph dingo.*

brown pigmented noses and eye rims. Generally, the ratio of height to body length is 4:5. The AD's wedge-shaped head is wider than its chest, allowing it to negotiate through small openings. AD leg length and moderate joint angulation maximize endurance while using minimum energy. Their front digits can grasp objects.

There are three ecotypes of AD. The "alpine" type has a heavily brushed tail and a seasonally thick double coat with neck ruff and breeching. The alpine type also has heavier bones and sometimes slightly shorter limbs than the other types, natural adaptations for cold mountain habitats. The "central" or "desert" ecotype has a shorter, harsh outer coat with a lighter seasonal undercoat, a medium-brushed tail, and long legs for long-distance plains travel. The third ecotype, referred to as "tropical" (forest adapted), has a short, crisp coat, no tail brush or undercoat, and appears finer built throughout than the other varieties.

BEHAVIOR

What sets the dingoes completely apart from domestic dogs is their niche: they are apex predators. The dingoes are the only dogs with the ability to fill the same role as the smaller wolves of Eurasia. Captive AD temperament varies widely, from extremely shy to friendly. They are not aggressive without a reason. The AD's high intelligence is legend. In captive conditions this cleverness leads to human amazement and frustration. Keeping a dingo safely captive takes ingenuity and intensive escape-proofing. Although ADs have mistakenly been labeled pack animals, in fact AD social behavior varies and is less structured than wolf families. ADs usually are loners or live as bonded pairs, sometimes with one or two young from a previous breeding season still accompanying them. They can form co-operative hunting groups when the prey is larger than individual AD body size.

FIGURE 4. *This adolescent wild male demonstrates dingo sure-footedness and agility in its head-first run down a cliff face.*

CULTURAL ASSOCIATIONS

When the first European explorers arrived in the 18th century at Sydney Cove on the east coast of Australia, they mentioned seeing the dogs with the aboriginal people. The earliest ships to enter the southern bays also recorded sighting yellow and black dogs with the native people. Although a few early reports state ADs were used to hunt large kangaroos, most say the tame ADs were mainly pets that assisted the people with finding small hidden prey. The people captured wild-born puppies out of the dens and raised them in the villages. Most of these went back to the wild after becoming sexually mature, and did not breed in captivity. The dingo is a major character of the aboriginal dreamings, the spiritual journeys of the people, and is involved in myths about the

FIGURE 5. *This 9 week old Australian dingo is determined to escape its pen, and already has the agility and intelligence to do so. It is squeezing its body through the small opening designed to admit a hand to undo the gate latch.*

FIGURE 6. *A dingo puppy about 7 weeks old looks out from the safety of a hollow log.*

origin of the aboriginal people. In some dreamings they see their ancestors in dingo form before becoming human.

CONSERVATION

In the past, the AD roamed over the entire continent of Australia, an area about equal to the size of the continental USA. Today the AD is found only in 3% of Australia, pushed into deserts and remote places by human persecution, and by the loss of habitat to agriculture and livestock production. There are no over-all population statistics for the wild AD. In many places it is classified as "vermin," subject to predator control measures such as poisoning, trapping and shooting. The populations most remote from developed areas are still pure, but hybridization with domestic dogs is rampant in several areas. Recent genetic research is helping to define the origin and population structure of ADs, and AD/domestic dog hybrids can now be distinguished by DNA tests. The hope is that as evidence accumulates that the AD is an irreplaceable ancient race of natural dog, efforts to conserve them as a pure subspecies will increase.

For more information: www.dingodiscovery.net/

ABORIGINAL LANDRACE DOGS

AFRICANIS: THE ABORIGINAL DOG OF SUB-EQUATORIAL AFRICA

BY JOHAN AND EDITH GALLANT

FIGURE 1. *Young men with their Africanis, probably gathering for a hunt, in KwaZulu Natal.*

Africanis is the umbrella name for all Southern Africa native dogs. It refers to Africa (the continent) and *Canis* (dog). Over the centuries they have been naturally shaped by Africa for Africa. They are part of local biodiversity and the cultural heritage of humankind. Africanis are only found in the tribal rural areas across the southern African subcontinent. They should not be confused with the multitude of mixed-type dogs roaming free in informal settlements and townships.

APPEARANCE

The beauty of this dog is embodied in the simplicity and functionality of its build. The Africanis is medium sized, slender, and well-muscled. They are usually 14–19 in. at the

FIGURE 2. *Africanis in a village in Mozambique.*

shoulders (50–60 cm). Agile and supple, they move in a very natural and easy manner, and can run at great speed. The Africanis has the stamina to trot for long distances on rough terrain and hilly environment, and in gallop it can reach great speed when required. When in good condition, the ribs are just visible. The head is cone shaped with expressive oval eyes. Eye rims, lips, and nose should be black. The ears may be erect, half erect, or drooping. The carriage of ears and tail is linked to the dog's awareness of its environment. These variable physical features are of no direct influence on the physical and mental well-being of the dog. The dog has a short double coat and is found in a wide range of colors, with or without white markings. A so-called ridge (hair growing in reverse direction down the spine) of varying form may be present.

FIGURE 3. *These Africanis are from Botswana. Note the ridge on the one lying down.*

ORIGIN OF THE AFRICANIS

Their African heritage goes back 7000 years, to the dogs which came with Neolithic herdsmen from the Middle East into the then dog virgin continent of Africa. Even before the time of the Egyptian dynasties, domestic dogs spread quickly along the Nile River. At the same time, seasonal migrations and trade took them deep into the Sahara and Sahel. Iron-using Bantu speaking people brought their domestic dogs along when, from about 200 AD, when they left the grasslands of Cameroon in a massive migration that eventually led to their settlement in Southern Africa.

FIGURE 4.
An Africanis dam nursing her young pups in KwaZulu Natal. The tips of her ears appear to be damaged, likely due to fly bites.

CHARACTERISTICS

Because the Africanis has for centuries roamed freely in and around rural settlements, it combines attachment to humans with a need for space. Traditionally it is always close to humans, other dogs, livestock, and domestic animals. It has a natural tendency to guard and protect livestock. The Africanis is well disposed without being obtrusive: a friendly dog, showing watchful territorial behavior. This dog displays unspoiled social canine behavior with complex facial expression and body language. Its nervous constitution is strong, with a high level of natural survival instinct demonstrated as caution and alertness.

THE AFRICANIS IN THE FUTURE

The Africanis as a primitive hound is guided by the instinct of subservience, the very drive that made its distant ancestors prime candidates for domestication. It is bound to its human partners and its territory. It will follow you for hours without being on a lead. From the moment we take the Africanis away from its natural habitat we are interfering with its future. On the other hand, its historic rural habitat is changing and shrinking at an alarming rate. To conserve the Africanis, they can be utilized as a real working dogs, such as flock guardians, search and rescue, or tracking dogs.

The vision of the Africanis Society of Southern Africa is to conserve the Africanis as a heterogeneous land race. For ages these dogs have shared the rural lives of Bantu and Khoisan people. They are part of their cultural and historical heritage. They do not need western style 'breed improvement.'

For more information: http://www.Africanis.co.za/index.htm
Gallant, Johan. 2002. *The Story of the African Dog.* U. of Natal Press, Pietermaritzburg.

CENTRAL AFRICAN DOG

BY JO THOMPSON, PhD

GEOGRAPHIC RANGE

FIGURE 1. *West Central Africa. The range of the Central African dog is shown in green.*

The Central African Dog is found in the forested landscape corresponding to the hydrographic low elevation basin that straddles the equator at the center of the African continent. Known as the Congo Basin, this geographic range encompasses the watershed dominated by the Congo River and its myriad of feeder tributaries; the ecosystem defined by old-growth, ancient, equatorial rainforest that includes those forested areas found in Equatorial Guinea, Gabon, Congo Republic, Democratic Republic of Congo, as well as southern Cameroon, southern Central African Republic, northern Angola, and southern South Sudan.

PHYSICAL DESCRIPTION

The Central African Dog is an elegant and agile hunting partner, both structurally sound and well-balanced. Typically they are 15–18 inches (38–46 cm) in height at withers where males reach the taller end of the range and females smaller but not dimorphic. Their weight reflects an active lifestyle and conditioning.

Their coats are short, giving the appearance of being painted-on, with no under coat. The hairs are like a soft brush; just stiff enough to give some protection when moving through the forest understory, keep dirt from clinging, and repel light rain/morning dew, yet yielding enough to give a uniform texture and the illusion of being soft.

They are easily recognized by the tail curl that may coil full circle or more. Their ears are erect, cupped for optimum sound reception, comparatively smaller than typical hound ears, and the sides of the ears curve frontward while the tip is forward of the base to give a slightly hooded appearance. Their expression is one of the most important hallmarks of the Central African Dog. That expression is created by the combination of eyes, wrinkled brow and cheeks, muzzle-cushioning and structural definition. The almond-shaped, obliquely set eyes are enveloped by dark pigmented rims giving the illusion of far-seeing, inquisitive, focused-concentration. The overall result is an expression

FIGURE 2. *Black, tan and white (tri-color variety).*

giving the appearance of a rather impenetrable, mysterious, thinking-mind highly adept at problem-solving drawn from the cumulative wisdom of the ancestors gained over thousands of years.

The skin is very pliant and allows for maneuvering when held within the jaws of an aggressor. The predominant coat color is based on a spectrum of red to fawn, with white on all four feet, point of chest, and tail tip. Additional colors include tricolor (black and red with white feet, chest, and tail-tip), black (with white feet, chest, and tail-tip), and more recently red (as above) with a black-brindle pattern over the red.

FIGURE 3. *A litter of puppies about 4 weeks old.*

BREEDING SEASON

The Central African Dog is found on both north and south latitudes radiating out from the equator. They are known to have an annual breeding season which corresponds to the timing of the rainy season and puts puppies on the ground during the dry season: north of the equator puppies are more prevalent around February while south of the equator they are more common around August. During the rainy season, forest-dependent people may rely more heavily on other sources of meat and will leave the
village to travel distances required to reach temporary fishing camps, sometimes taking a few dogs with them. During the dry season, hunts are more successful and forest meat (primarily ungulates and monkeys) is more abundant.

FIGURE 4. *This female illustrates the short coat and hooded ears of the Central African dog.*

FIGURE 5. *This group of Central African dogs looks very well cared for.*

CULTURAL ASSOCIATION

The people, who know only traditional forest-based livelihoods, are dependent on hunting large-bodied mammals for survival. They are a proud hunting culture constructed around the desire for a meat-based diet. Thus, their lives in the ancient equatorial primary rainforest require a specific arsenal of traditional hunting tools. The dogs are the property of the hunter; a part of the forest tool-kit along with bows, arrows, spears, and nets made from local plant fibers. Forest meat is the primary source of animal protein essential for household subsistence for these forest-dwelling people, different from the diets of the river people who subsist primarily by fishing and trading. Duikers (forest antelopes) are among those species most preferred by the forest hunters. It is working the hunt for these hoofed mammals where the Central African Dog is a specialist.

TRADITIONAL USES OF THE DOGS

The dogs are first and foremost a part of the hunting effort. Secondarily they offer some companionship but that is not openly acknowledged. Isolated principally by the geography of the Congo Basin and traditional technology, the Central African Dog population established in a niche as a specialized tool used by humans dependent on that ancient forest for their livelihoods in the traditional net-hunt targeting large-bodied, hoofed animals.

FIGURE 7. *Dogs at home.*

For more information:
http://www.rvwbasenjiclub.org/LukuruProjectThree.html
https://www.basenji.org/african/project.htm

THE CANAAN DOG OF ISRAEL

BY MYRNA SHIBBOLETH

FIGURE 1. *These Canaan dogs are part of the captive breeding program of Shaar Hagai kennel in Israel. They descended directly from Bedouin stock.*

GEOGRAPHIC RANGE

The dog that is known as the Canaan Dog is in fact a type of aboriginal dog that is found through the Middle East, and has been known in the area for thousands of years with little or no changes. It is ideally suited to live in this primarily arid part of the world, with scarce water, very hard, rocky, and hilly terrain, and limited food sources.

FIGURE 2. *This Canaan dog has the thick double winter coat that insulates them against the desert's cold nights.*

FIGURE 3. *The Canaan dog, like all natural dogs, has an efficient trotting gait that can be maintained for hours across all types of terrain.*

The Canaan Dog is the specific type that has developed in what was biblically called the "Land of Canaan", or what is present day Israel. Israel is the only country in the area that has made an effort to preserve these dogs, and has recognized them as the national breed (today recognized throughout the world). However, although almost all the basic stock has come from dogs found within the borders of Israel, there are occasionally additions, when there is an opportunity, of dogs from Jordan or Egypt that fit the desired type. Due to the pressures of civilization, and the disappearance of their natural niche, the dogs are today rapidly disappearing as free living animals.

PHYSICAL DESCRIPTION

The Canaan dog is medium sized and athletic, from 19 to 23.4 in. (50 to 60 cm.) in height at the withers, with the females on the smaller side and the males toward the top of the scale, and with medium bone. He is square in build, with a wedge shaped head, erect ears that are set obliquely, almond shaped dark eyes, all of which give him his alert and appealing expression. He has a double coat, with a short to medium strong and weather resistant outer coat and very thick woolly undercoat according to the season. His tail is carried over his back, and is very much an instrument of communication, conveying his feelings and moods. Colors are quite varied, from sand to red, or black, with white trim, or white with black or red to cream spotting. It is very important that everything conveys functionality and fitness. Canaans are quite muscular dogs. Movement is very light and agile, and they are easily able to jump and maneuver.

FIGURE 4. *This litter of Canaan puppies was found outside a Bedouin camp.*

BREEDING SEASON

Breeding season is early spring and early fall, although it is not rare for a bitch to come in season once a year in the spring. It is also not rare if there has been a period of drought for the females to not come in season or not accept the male.

CULTURAL ASSOCIATION

The Canaan has been associated for many generations with the Bedouin of the area. They serve as a warning system for anyone or anything that nears the tents or the flock, and as protection for the flocks. They will bark in warning, but they will also attack animals that are endangering the flock, and are known to have attacked wolves and hyenas. The Bedouin very much value a working dog. They do not breed them, but know where the bitches den, and when they need a new working dog, will catch a male puppy and raise it in the camp. They are not willing to sell or trade a dog that has been a good working dog for them, and they will "honor" old dogs that have served

FIGURE 5. *Canaan dogs are extremely hardy and athletic. This one is running the dunes near the Dead Sea.*

them well by letting them live out their lives in the camp. They are not pets, and the men are rarely able to come near or touch them, but the children of the camp have a different relationship and often can catch and play with the dogs, and the women often give them additional food. Although there may be a number of families and flocks in the same encampment, each dog knows to which family and herd he belongs and stays with them.

TRADITIONAL USES

The Canaan has always been used as a guard and flock protection dog. His extremely keen and well developed senses and his adaptation to the difficult climate and terrain made him a very useful dog for the military at the inception of the state of Israel, when they were used frequently for a variety of guard and patrol duties and nose work.

For more information: http://canaandogs.info/

THE FORMOSAN MOUNTAIN DOG

BY MING-NAN CHEN

MYTHICAL ORIGIN OF THE FORMOSAN MOUNTAIN DOG

In the mythology of Atayal aboriginal people regarding the origin of Formosan Mountain Dog, human did not keep dogs at ancient time; all dogs was wild animal, whom lived in wild condition and lived in a location be named "wild dog valley". This animal was horrible; it not only bit human, but also ate human. They hunted other animals include wild boar, deer, and serow (a goat-like animal) in large numbers. Atayal people's ancestor thought "if we can domesticate dogs and let dogs help people hunting, would not it be nice? But what a ferocious those dog! How can we capture and domesticate?" Finally, they dispatched three young courageous men to perform the task. The fastest man carried traditional ethnobotanical food, millet mochi (sticky millet cake, very sticky). When adult wild dogs went away the from their cave, two men captured two puppies to place each behind the mesh bag respectively on the sly and escaped quickly. The third man who had mochi brings up the rear and ran in different directions continuing to throw millet mochi on the ground. When the wild dogs heard puppy sounds, they were catching up those puppy thieves. More and more wild dogs chased the

FIGURE 1. *A free-ranging Formosan Mountain Dog in a traditional village.*

three young people. Because those millet mochi had been dipped in boar grease, wild dogs immediately snatched mochis. Fortunately, after they ate the millet mochi it stuck on their teeth and they could not bite. In order to shake off the mouth of millet mochi, they walked around in circles, seeming to have forgotten to rescue the puppies. This is the origin mythology of Formosan mountain dog. Those two wild dog puppies become the first generation hunting dogs, Atayal ancestor kept, bred and handed down to descendants ...

HABITAT

The Formosan Mountain Dog (FMD), also called the Taiwan mountain dog, originated from the high mountains of the island of Taiwan. Taiwan was formerly called Formosa, a Portuguese historical name meaning beautiful island. There are 286 mountain peaks over 9,800 ft. (3,000 m) above sea level in the small island area of 22,500 sq. mi. (36,000 sq. km.), so there is a huge variation in elevation from steep high mountains to deep valleys. Forests cover 70% of the island's mountains. The lowland subtropical climate is heavily influenced by the ocean and the East Asian monsoon, which create high humidity and ample precipitation. FMD were formerly present in the Taiwanese high mountain area including the following ranges: Snow Mountain, Central Mountain, Ali Mountain, Morrison Mountain, and Coastal Mountain.

CULTURAL ASSOCIATION

Since ancient times, the dog was kept by Austronesian-speaking people who migrated to Taiwan about 6000 years ago. Human and dog cohabitated together in partnership, surviving the challenging environment for thousands of years.

Hunting was very important in the traditional Taiwanese

FIGURE 2. *FMDs regularly climb trees with limbs that will hold their weight.*

Austronesian-speaking aboriginal cultures. It was vital to human survival. The FMD has a strong desire to hunt and was an essential hunting partner in the forests. In addition when different tribal hunting territories overlapped, causing inter-tribe conflicts, the FMD was an important guard and alert dog when outsiders and strangers attempted to invade a village. For the above reasons, this dog possesses sensitive scent, vision, and hearing, and it is sharp, alert and courageous to protect and attack when guarding.

FIGURE 3. *This FMD is climbing an almost vertical hillside, demonstrating their power and fearless nature.*

DESCRIPTION

The FMD is a medium dog, weighing between 30-55 lb. (15-25 kg). Their shoulder height is about 20 in. (50 cm) for males and about 17 in. (45 cm) for females, plus or minus 2 in. (5 cm). The typical appearance includes a streamlined body with a deep chest and narrow waist. The triangular shaped head has a broad skull, medium snout-length, erect ears, and orange, yellow or dark brown almond shaped eyes. The tail is curled or sickle shaped, giving the body a good sense of balance and stability. The claws of the flexible fore paws point downwards, permitting a strong grip and excellent digging ability. The moderately curved and well-developed hock of the hind legs provides strong jumping ability. Coat color is unimportant and many be black, brindle, yellow, fawn, red, sesame, cream, white-black, or chocolate, etc.

The important key point is hair quality. The coat must be a double, made up of inner fine hairs (to keep it warm and waterproof) and coarse outer hairs (to provide protection against burrs in the bush and to shed dirt). The FMD puppy matures faster than most dogs so many behaviors appear earlier, especially state of mind. The

FIGURE 4.

A good example of the red coat variety showing the lighter areas on cheeks, legs, belly, and the underside of the tail which in the black and tan color becomes the tan. This pattern, a type of counter shading, is common in primitive dogs and wild canids.

FIGURE 5.
This litter of FMD puppies, about 5 weeks old, were born in this den excavated by their dam. The tunnel to the nest is as long as a man's arm.

adult temperament is an uncultivated type of personality. FMDs are strong-willed, independent thinkers with the following traits: Full of team spirit when group working; suspiciousness of strangers; strongly territorial; not prone to random barking. They have excellent comprehension and memories. Loyal and affectionate to their families, they accept responsibility and seem to have a human-like spirit. At the same time, because this dog is a kind of semi-wild animal that is allowed to range free, it retains innate instincts such as strong ability to identify direction, an accurate sense of time, and even seemingly telepathic ability. Those traits reflect the dogs' history of mountain life.

For more information: Formosan Mountain Dog Conservation Center: http://www.dogs.com.tw/eg/indexeg.htm

INDIAN NATIVE DOG

BY RAJASHREE KALAP AND GAUTAM DAS

FIGURE 1. *This Indog was seen in western India at the Konkan coast. It is a fine example of the slender ecomorph common in the plains and desert regions.*

The 'Indian Native Dog' or INDog, commonly also called the Indian Pariah Dog (or 'Pi dog') is possibly the original type of domestic dog in the Indian subcontinent. It is an autochthonous landrace of early-domesticate dog.

While in the distant past it may have been the predominant village dog of India, today it is not. Apart from the INDog, India has a number of indigenous breeds developed from the Indog, corresponding to different geographical zones and landscapes: Mainly livestock-guarding breeds and sight-hounds. However, the INDog has a wider range than any other landrace or breed of the country.

DESCRIPTION

The INDog appearance conforms to the dingo-pariah type seen in most of the village dogs of Asia and Africa. It is medium-sized, ranging in height from 18–25 inches (46–63.2 cm) at the withers, and in weight from 30–66 lbs. (15–30 kg). The coat is short. It usually is a double with seasonal insulating underfur, but a single coat is seen in some

FIGURE 2. *The smaller eastern tropical ecomorph Indog from Nameri Assam.*

areas. The commonest colors are brown (shades varying from beige to dark red), with or without white markings on the feet, legs and tail tip. White with brown or black patches/spots coats are frequently seen in some areas. Tricolors (black with tan points and white markings) are common in some regions. Brindled coats and black saddles are rare and may be an indication of mixed ancestry, but this needs genetic research. The head is medium-sized and wedge-shaped, with the pointed muzzle of equal or slightly greater length than the skull. The eyes are almond-shaped and typically dark brown. The ears are held erect and are pointed at the tips with a broad base, set low on the head. The tail is always curved and sometimes tightly curled.

There are regional variations within the basic type, with the INDogs of central India on average taller than in other areas; those of eastern India a little smaller with smaller ears; and those of the north-eastern valleys and lower hills somewhat stockier with a slightly shorter muzzle than the peninsular Indian and northern Indian populations.

BREEDING SEASON

Like other aboriginal dogs, INDogs come into estrus once a year. The seasonal, synchronized breeding cycle is usually in the cooler monsoon rainy season (July through October) in tropical peninsular India. In the sub-tropical northern Indian plains and Himalayan foothills, the breeding season is from end-September to end-October, with litters born in the cold winter months, mainly in December and January.

FIGURE 3. *An Indog dam cares for her pups on the beach in Sri Lanka.*

TRADITIONAL ROLES

Whether an INDog is owned or ownerless depends on the prosperity of the human community in which it lives. In impoverished societies inhabiting small hamlets, the dogs are almost invariably all owned, since such societies cannot afford to waste food and do not generate enough garbage to sustain a scavenging dog population. Examples are the small villages in forested areas of Odisha, eastern India. On the other hand, villages and towns with higher levels of consumption and edible waste support varying numbers of ownerless scavenging dogs.

USES AND CULTURAL ASSOCIATIONS

Dogs do not have high status in mainstream Indian cultures, exceptions being some traditional aboriginal societies. However they are not a taboo animal and villagers acknowledge their usefulness. Owned INDogs are mostly working dogs used as flock or property guardians. An older niche is that of hunting partner, especially in aboriginal communities such as of the Gonds of central India and the Santhals of the Chota Nagpur Plateau in east-central India. Hunting is now illegal under India's Wildlife Protection Act, but continues in areas with weak law enforcement. Interestingly, INDogs used for hunting wildlife will usually not display any predatory behavior towards village livestock or poultry. Some village dogs are companion dogs and treated with considerable affection. Owned dogs are left to roam and breed freely. In many, though not all, rural communities, a pet dog may be discouraged from entering the house and sleeps outside.

FIGURE 4. *This Indog, seen in the plains area of Assam, is an uncommon cream color.*

INDogs very rarely live in the truly feral niche with no direct interaction with people. The ownerless dogs are scavengers dependent on humans for food and reside in or very near the villages. There are some reports of feral packs that survive by hunting wildlife, but this is the rare exception rather than the norm.

Local names for the Indog vary with each regional language, but almost all of them mean 'native', for example *Desi kutta* (Hindi). The Marathi name is *Gauthi kutra* (village dog).

GEOGRAPHICAL RANGE

INDogs are found in plains areas and the lowest mountain ranges, typically in agricultural areas and around human settlements near forests, and along the coasts. INDogs are usually absent in open grasslands regions because local dogs kept there are often sight-hound breeds or sight-hound mixes, developed locally for chase hunting. Mountainous regions also have locally developed working breeds, like the livestock guarding breeds of the Himalayas, and the Dhangar dogs of the Western Ghats and Deccan Plateau. Industrialized towns and cities are excluded from the range of pure INDogs and other Indian aboriginal dogs, replaced by free-ranging mixes with non-native breeds.

TEMPERAMENT

INDogs are very alert, territorial, and protective of their pack/family. They are cautious and suspicious of strangers, and extremely quick learners, qualities essential for survival in a free-ranging niche. Their prey drive can be high, but varies among populations according to their environment and personal history. INDog attacks on livestock are rare since this trait is strongly discouraged by humans. Dogs that attack poultry or domestic animals are almost always killed. However, INDogs which show no prey drive towards poultry resident in their own home village may prey upon poultry from another village

FIGURE 5. *This handsome older red ginger male Indog photographed in Sawra.*

that wanders onto 'common' farmlands between villages. There is no doubt that they recognize individual livestock belonging to their own 'home' area.

DIET

INDogs can digest both cereal and dairy products well. Like all dogs, they need animal protein in order to thrive, but their requirement is small. In rural households that 'own' or feed dogs, unleavened bread made from wheat or millet, or cooked rice, are the normal staples fed, supplemented with a little dairy product, such as milk, yogurt or whey, which provides some animal protein.

HEALTH

No inherited health issues have been recorded. However, in the village dog niche they rarely have a long lifespan. Only well-cared for individuals reach the age of ten years at most, so geriatric problems are not prevalent.

HISTORICAL ASSOCIATIONS

The curled tail of native dogs is a well-known feature and is referred to in ancient literature including the *Panchatantra* fables. The INDog type, with its erect ears and curled tail, is easily recognizable in ancient art, for example, the cave art of Bhimbetka in central India (from the Mesolithic/prehistoric period) and the Barhut Stupa sculpted railing (about 200 BC). Many dog remains have been found in archaeological sites of the Bronze Age Harappan civilization (3300–1300 BC) and the skulls are described as matching the type of modern village dogs, indicating the INDog has not changed significantly in thousands of years.

For more information: www.indog.co.in/
indianpariahdog.blogspot.com/
https://www.facebook.com/groups/254481394732431/

THE INUIT DOG
Canis familiaris var. borealis

BY SUSAN HAMILTON

FIGURE 1. *Inuit dogs working in a traditional "fan hitch" style pulling formation.*

HISTORY AND GEOGRAPHIC RANGE

The Inuit Dog dates back to the paleo Inuit culture of about 4,000 BP (Before Present), and originally accompanied Arctic-adapted humans across the Bering Strait, eventually migrating with them to Greenland. However, the earliest identified sled runners and harness material date to the Thule culture of about 800 BP. Between these two time periods the Thule cultures designated Independence I, Pre-Dorset, Independence II and Dorset used this aboriginal dog as a hunting partner and pack animal, not a sledge dog.

APPEARANCE, REPRODUCTIVE PROFILE AND BEHAVIOR

There is no written "standard" as there are for modern registered breeds. The breeding of Inuit Dogs was based on stamina and performance, not outward appearance. Lifestyle and harsh polar conditions shaped it through survival of the fittest and most

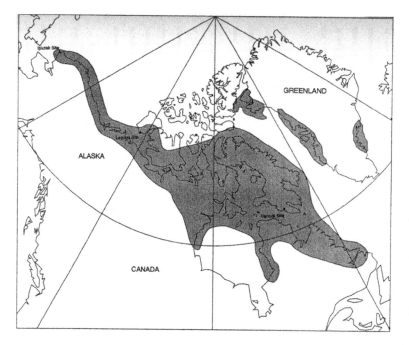

FIGURE 2.
Inuit Migration from Alaska across Canada to Greenland. The Greenland Dog and Canadian Inuit Dog have been proven by DNA analysis to be the same landrace.[1]

functional. There are generalizations that can be made:
- A hallmark trait, ears unfold earlier than most dogs, at about 3-4 days, becoming upright and remain so for life unless injured.
- Coat has two layers: a thick, insulating undercoat covered by longer, harsher guard hairs. Guard hair surrounding the neck and shoulders is longest, particularly in males. The extra-long hairs, 5-8 in. (12.7-20.3 cm) in length, make the males appear bigger to other males, an advantage when challenging each other for breeding rights.
- Coat colors range from all white to nearly all black or dark brown. There are shades of red and agouti (red mixed with dark brown or black) as well. White dogs can have spots of color, especially on the head.
- The eyes can be any shade of amber to brown, but never blue.
- Tails typically curl loosely over the back and then lay against one side of the body.
- The general impression of the overall body style is a robust, sturdily built dog.
- Males are usually significantly larger than females. According to Ian Kenneth MacRury in his masters thesis *The Inuit Dog: Its Provenance, Environment and History*, approximate weights and heights are as follows:

	Male	Female
Weight	84.9 lbs (38.5 kg)	67.5 lbs (30.56 kg)
Height	24.3 in (61.7 cm)	22.4 in (56.9kg)

FIGURE 3.
This individual's name has been preserved: Manake. This was taken at Bernard Harbour during the Northwest Territories Canadian Arctic Expedition of 1915.

FIGURE 4.
The dog is lying down while the woman starts to unpack its burden after a 10 mi (16 km) hike to Sapotit from the coast at Pelly Bay in June 1963.

TRADITIONAL USE AND LIFE

Inuit Dogs perform many essential skills: scent location of seal breathing holes and birthing lairs, tracking and detaining prey, alerting people to the presence of bears, hauling heavily laden *qamutiq* (sledges) over snow and ice, and carrying belongings on their backs in summer. The Inuit dog is acknowledged as the principal reason, right up to the mid-twentieth century, that the ancestors of today's Inuit survived. These dogs have a legendary resilience to hardships, being physiologically and anatomically adapted to survive and work at extremely low temperatures, even when fed on an irregular basis during periods of starvation when food was scarce for the hunting society.

Recently, the lives of these nomadic hunters changed dramatically. Inuit society has transitioned from seasonal hunting and trapping to living in permanent communities. This conversion accelerated beginning in the 1950s as a result of the "Dog Slaughter,"[2] carried out by the Canadian Government in the 1950-1970s. The introduction of the snowmobile had permitted more rapid transit for Inuit who commute regularly to jobs and fewer dogs were needed.

Traditionally, these dogs ranged free, allowed to roam outpost camps socializing with humans and their own species. When not working in some capacity, the dogs foraged for their own food. Their strong predatory behavior is thus a result both of their use as human hunting partners and their own need to survive when not being fed directly by humans.

THREATS TO SURVIVAL AND FUTURE

The presence of non-indigenous dogs is now ubiquitous throughout the Canadian North and currently there are no regulatory mechanisms in place to control this influx, either by outright banning or by allowing only altered dogs to accompany their owners north. In Greenland, there is a law forbidding non-indigenous breeds entering regions where sled dogs are kept, as a measure to keep the Greenland Inuit Dog uncontaminated. However, enforcement has not been strict. In May 2000, the one year-old government of the Canadian Arctic Nunavut Territory honored the Inuit Dog – not the seal, caribou, musk ox or even the iconic polar bear — as its official territorial animal.

FIGURE 5. *This photo of an Inuit woman and her packing dog was taken in the Northwest Territories between 1900 and 1929.*

FIGURE 6. *An Inuit dog with very young pups, taken at Bernard Harbour, Northwest Territories, 1915. Even if provided a small shelter as most are today, Inuit puppies have always had to survive extremely harsh conditions*

Despite the socio-economic challenges that continue to keep their future uncertain, this aboriginal dog endures in regions of arctic Canada and the northern and eastern districts of Greenland. While largely gone from Alaska and the western Canadian Arctic, some Inuit Dogs are present there for recreational sledding and tourism. Recent genetic results indicate the Inuit Dog remains a pure evolutionary line, separated from other dogs for at least 2,000 years.[3]

Today few Inuit can follow in the footsteps of their Elders and live a traditional lifestyle using dogs. However, many Inuit feel strongly about their culture and, along with a few dedicated non-Inuit, keep and use these dogs. Thus, traditional Inuit Dogs can still be found in their native habitat, used for harvesting meat for people and dogs, as well as sport hunting, eco-tourism and expedition travel.

For more details about the Inuit Dog, visit "Defining the Inuit Dog" at The Fan Hitch, Website and Journal of the Inuit Sled Dog, <http://thefanhitch.org/theISD/Introduction.html>

FIGURE 7. *These Inuit pups are eating strips of raw meat.*

NOTES

1. Population Genetic Analyses of the Greenland dog and Canadian Inuit dog, May 2005 by Dr. Hanne Friis Andersen, Royal Veterinary and Agricultural University, Frederiksberg, Denmark; S K Brown, C M Darwent, E J Wictum and B N Sacks. 2015. Using multiple markers to elucidate the ancient, historical and modern relationships among North American Arctic dog breeds. *Heredity* advance online publication 24 June 2015; doi: 10.1038/hdy.2015.49.

2. Final Report: Allegations concerning the Slaughter of Inuit Sled Dogs in Nunavik (1950-1970); The Fan Hitch, http://thefanhitch.org/officialreports/Final%20Report.pdf; Qikiqtani Truth Commission (Nunavut), Achieving Saimaqatigiingniq, The Fan Hitch, http://thefanhitch.org/officialreports/QTC%20Final%20Report%20EN%2013%20Oct%202010.pdf.

3. Antiquity of the Inuit Sled Dog Supported by Recent Ancient DNA Studies; Sarah K. Brown, PhD; Anthropology and Veterinary Genetics, University of California, Davis.

ALBUM OF DOGS

FIGURE 1. *Canada, Northwest Territory, Inuit 1991.*

FIGURE 2. *Canada, Baffin Bay, High Arctic. Inuit dogs at far edge of floe.*

FIGURE 3. *Africa. Basenji.*

FIGURE 4.
Formosa Mountain Dog Perching.

FIGURE 5. *Formosa Mountain Dog with her den and puppies.*

FIGURE 6.
Indonesia (Bali).

FIGURE 7. *Indonesia, Toraja, Sulawesi.*

FIGURE 8. *Indonesia (Flores). Puppies.*

FIGURE 9. *Indonesia (Sulawesi). Note the ridge similar to those in Africanis.*

FIGURE 10. *Indonesia (Sulawesi).*

FIGURE 11. *Indonesia (Sulawesi).*

FIGURE 12. *Indonesia (Borneo)*

FIGURE 13. *Korea (Jeju Island). Village dog.*

FIGURE 14. *Nepal (Kathmandu).*

FIGURE 15. *Korea (Jeju Island). Village dog*

FIGURE 16. *Mongolian village dogs.*

FIGURE 17. *Malaysia (Tioman Island).*

FIGURE 18. *Philippines. Mountain Tiger Dogs.*

FIGURE 19. *Papua New Guinea.*

FIGURE 20. *Philippines. Mountain Tiger Dog*

FIGURE 21. *West Siberian Lakia.*

FIGURE CREDITS

Front Cover: Kerrie Goodchild
Frontispiece: shutterstock_140527951/FiledIMAGE

PART I.

Frontispiece: Lynn Jackson

Chapter 1: 1.1. Wiki Media Commons/Gustav Doré 1867; 1.2. Thinkstock_120056076/ Images in the Wild; 1.3. Thinkstock_174105709/ hammett79

Chapter 2: 2.1. Janice Koler-Matznick and Karen Adair; 2.2. Janice Koler-Matznick; 2.3 Janice Koler-Matznick; 2.4. Janice Koler-Matznick; 2.5. Janice Koler-Matznick; 2.6. iStock_000008519161/Mylifeiscamp

Chapter 3: 3.1. I. Lehr Brisbin, Jr.; 3.2. Monty Sloan, Wolf Park; 3.3. Wolf paws: Fotolia_65475738/mariemilyphotos, dingo paws: shutterstock_64983883/ Nicholas Lee; 3.4. Karen Adair/Janice Koler-Matznick/adapted from sjulienphoto/iStock_000002992587 and MoMorad/iStock_17787903; 3.5. Adapted from Ovodov N. D., et al. 2011 doi:10.1371/journal.pone.0022821; 3.6. Shutterstock_264393941/odd-add

Chapter 4: 4.1. shutterstock_38177203/Mopic; 4.2. Janice Koler-Matznick, adapted from Wayne & Ostrander 1999 (Fig 3., P. 250); 4.3 Wiki Commons/Darwin's Monkey 2008/ HugoRheinholdApeWithSkull.DarwinMonkey; 4.4. Wiki Commons/Thomas Lersch; 4.5. Dr. Rudi Turner/The Interactive Fly

Chapter 5: 5.1. Janice Koler-Matznick; 5.2. shutterstock_129179390/Jeannette Katzir Photog; 5.3. shutterstock_248868271/Michal Ninger; 5.4. U.S. Forest Service/Yellowstone National Park; 5.5 Patricia McConnell; 5.6. Coyote: ThinkstockPhotos-185108303/Paul Tessier, Jackal: iStock_000056130798/Chatterer; 5.7. Johan and Edith Gallant/Africanis Society; 5.8. shutterstock_269393150Noppanun K; 5.9. Rajashree Kalap/Indog Project; 5.10a, b, c. Sunil Pal; 5.11. Thinkstock_178034293/Sabirmallik; 5.12. Johan and Edith Gallant/Africanis Society; 5.13. shutterstock_220910986/Vladimir Kogan Michael; 5.14. Janice Koler-Matznick; 5.15. Thomas Newsome; 5.16a, b. Lyn Watson/Australian Native Dog Foundation; 5.17. shutterstock_138115103/Eder

Chapter 6: 6.1. Janice Koler-Matznick/Karen Adair; 6.2. Lynn Jackson; 6.3. iStock_000020497771/Angelika Stern; 6.4. shutterstock_218265754/Nopparatz; 6.5. iStock_000043413624/prwstd; 6.6. shutterstock_278937779/Jamie Hall ; 6.7 shutterstock_38838163.wildlywise; 6.8. shutterstock_198607/Sakala; 6.9. Lyn Watson; 6.10. Janice Koler-Matznick, adapted from Wiki Commons/Maximilian Dörrbecker (Chumwa)

Chapter 7: 7.1. shutterstock_113086033/Humming Bird Art; 7.2. shutterstock_329873966/ HacK-LeR; 7.3. Patricia Lee; 7.4. Flickr/RoyLathwell; 7.5 Gabriel Bennett

PART II.

Frontispiece: Flickr/Paolicchio

Introduction: 1. Michael Whitesell; 2. iStock_000057294084/pixelfusion3d; 3. shutterstock_130537721/pitaya

Australian Dingo: 1. iStock_000024949581_JohnCarnemolla; 2. Lyn Watson; 3. shutterstock_64983883/Nicholas Lee; 4. Kerrie Goodchild; 5. Lyn Watson; 6. Lyn Watson

New Guinea Dingo: 1. © Clifford B. Frith; 2. Win Waringi; 3. Susan Bulmer; 4. Tim Flannery; 5. Susan Hendler

Africanis: All figures Johan and Edith Gallant

Canaan: All figures Myrna Shiboleth

Central African Dog: All figures Jo Thompson

Formosan Mountain Dog: 1. Ming-Nan Chen; 2. David Liu; 3. Chuan-Shun Lin; 4. Ming-Nan Chen; 5. Ming-Nan Chen

Indog: 1. Rajashree Kalap; 2. Rajashree Kalap; 3. Stephan Gillmeier; 4. Kiran Kalap; 5. Rajashree Kalap

Inuit Dog: 1. Corel Arctic Sledding; 2. Susan Hamilton; 3. Canadian Museum of Civilization archive CMC50955/G. H. Wilkins; 4. NWT Archives/Douglas Wilkinson fonds/N-1979-051: item 1812; 5. Library of Congress LOT 11453-1, no. 53 [P&P]/ Frank and Frances Carpenter collection; 6. Canadian Museum of Civilization archive CMC51568/G. H. Wilkins; 7. Susan Hamilton

Album: 1. Cees Andresen; 2. Susan Hamilton; 3. iStock_000043742872/Farinosa; 4. Ming-Nan Chen; 5. Ming-Nan Chen; 6. iStock_000032963490/amyjosmile; 7. Pepper Trail; 8. iStock_000051455856/Goddard Photography; 9. Gerben Groustra; 10. Gerben Groustra; 11. Gerben Groustra; 12.Wikicommons; 13. Bruce Campbell; 14. iStock_000021655250/EthanTremblay; 15. Bruce Campbell; 16. Neha Arora; 17. Barry Drew; 18. Tom Asmus; 19. Mike Wilangue; 20. Thom Asmus; 21. Thinkstock_482827456/Didi Lavchieva;

Back Cover: Lyn Watson

BIBLIOGRAPHY

Acland, G. M. and E. A. Ostrander. 2003. Population genetics: the dog that came in from the cold. *Nature* 90(3): 201-202.

Aggarwal, R. K., J. Ramadevi and L. Singh. 2003. Ancient origin and evolution of the Indian wolf: evidence from mitochondrial DNA typing of wolves from Trans-Himalayan region and Pennisular India. *Genome Biol.* 2003, 4:P6: http://genomebiology.com/2003/4/6/P6

Aggarwal, R. K., et al. 2007. Mitochondrial DNA coding region sequences support the phylogenetic distinction of two Indian wolf species. *J Zool Syst Evol Res* 45(2): 163-172.

Akst, J. 2010. Surprising mtDNA diversity. http://the-scientist.com/blog/print/57199/ (Accessed 02/15/2015).

Albert, F. W., et al. 2012. A comparison of brain gene expression levels in domesticated and wild animals. *PLoS gen.* 8(9): e1002962. Doi:10.1371/journal.pgen.1002962

Allen, B. L. 2010. Skin and bone: observations of dingo scavenging during a chronic food shortage. *Aust Mamm* 32(2): 207-208.

Allen, B. L. 2012. Do desert dingoes drink daily? Visitation rates at remote waterpoints in the Strzelecki Desert. *Aust Mamm* 34(2): 251-256.

Allen, B. L., M. Goullet, L. R. Allen, A. Lisle, and L. K. P. Leung, (2013). Dingoes at the doorstep: preliminary data on the ecology of dingoes in urban areas. *Landscape and Urban Planning* 119: 131-135.

Amarasekare, P. and R. M. Nisbet, 2001. Spatial heterogeneity, source-sink dynamics, and the local coexistence of competing species. *Am Nat* 158(6): 572-584.

Amstrup, S. C. 2003. The Polar Bear-*Ursus maritimus*: Biology, Management, and Conservation. Pp. 587-610 in G. A. Feldhamer, B. C. Thompson, and J. A. Chapman (eds.), *Wild Mammals of North America*, 2nd ed. John Hopkins U Press, Baltimore, Maryland.

Andelt, W. F. 1985. Behavioral ecology of coyotes in south Texas. *Wildlife Monographs No. 94*: 3-45.

Anderson, A. J. 1981. Pre-European hunting dogs in the South Island, New Zealand. *New Zeal J Arch* 3: 15-20.

Anderson, A. J. 1990. Kuri. Pp. 281-287 in C. M. King (ed.), *The Handbook of New Zealand Mammals*. Oxford Univ. Press, Melbourne.

Anonymous. 2012. Sanitation role of Indian street dogs quantified. *Anim People* 21(7): 18.

Anyonge, W. and A. Baker. 2006. Craniofacial morphology and feeding behavior in *Canis dirus*, the extinct Pleistocene dire wolf. *J Zoo* 269(3): 309-316. Ardalan, A., et al. 2011. Comprehensive study of mtDNA among Southwest Asian dogs contradicts independent domestication of wolf, but implies dog-wolf hybridization. *Ecol Evol* 1(3): 373-385.

Arendt, M., et al. 2014. Amylase activity is associated with *AMY2B* copy numbers in dog: implications for dog domestication, diet and diabetes. *Anim Genet* Doi: 10.1111/age.12179

Arnason, U, A. Gullberg, A. Janke and M. Kullberg. 2007. Mitogenomic analyses of caniform relationships. *Mol Phylogenet Evol* 45(3): 863-874.

Arnold, M. L. 1997. *Natural Hybridization and Evolution*. Oxford Univ. Press, NY.

Aronoff, J., B. A. Woike, and L. M. Hyman. 1992. Which are the stimuli in facial displays of anger and happiness? Configurational bases of emotion recognition. *J of Pers and Soc Psych* 62(6): 1050-1066.

Aschoff, J. 1966. Circadian activity pattern with two peaks. *Ecology:* 657-662.

Ashenafi, Z. T., et al. 2005. Behaviour and ecology of the Ethiopian wolf (*Canis simensis*) in a human-dominated landscape outside protected areas. *Anim Conserv* 8: 113-121.

Atickem, A., A. Bekele and S. D. Williams. 2010. Competition between domestic dogs and Ethiopian wolf (*Canis simensis*) in the Bale Mountains National Park, Ethiopia. *Afr J Ecol* 48 (2): 401-407.

Atkins, D. L. and L. S. Dillon. 1971. Evolution of the cerebellum in the genus *Canis*. *J Mamm* 52(1): 96-107.

Auersperg, A. M. I., G. K. Gajdon and A. M. P. von Bayern. 2011. A new approach to comparing problem solving, flexibility and innovation. *Commun Integrative Biol* 5(2): 140-145.

Auersperg, A. M. I., et al. 2011. Flexibility in problem solving and tool use of kea and New Caledonian crows in a multi access box paradigm. *PLoS One*: 6x20231 PMID: 21687666

Avise, J.C., D. Walker, D. and G. C. Johns, 1998. Speciation durations and Pleistocene effects on vertebrate phylogeography. *Proc Royal Soc London B: Biol* Sci 265(1407): 1707-1712.

Axelsson, E., et al. 2013. The genomic signature of dog domestication reveals adaptation to a starch-rich diet. *Nature* 495(7441): 360-364.

Baker, R. J. and R. D. Bradley. 2006. Speciation in mammals and the genetic species concept. *J Mammal* 87(4): 643-662.

Bandelt, H-J. 2007. Clock debate: when times are a-changin': time dependency of molecular rate estimates: tempest in a teacup. *Heredity* 100(1): 1-2.

Bardeleben, C., R. L. Moore, and R. K. Wayne. 2005. Isolation and molecular evolution of the Selenocysteine tRNA (*Cf* TRSP) and RNase P RNA (Cf RPPHI) genes in the dog family, Canidae. *Mol Biol Evol* 22(2): 347-359.

Bardelen, C., R. L. Moore, and R. K. Wayne. 2005. A molecular phylogeny of the Canidae based on six nuclear loci. *Mol Phylogenet Evol* 37: 815-831.

Beach, F. A. and B. J. LeBoeuf. 1967. Coital behavior in dogs. *PNAS* 61: 442-446.

Beck, A. 1973. *The Ecology of Stray Dogs: A Study of Free-Ranging Urban Animals*. Purdue University Press, West Lafayette, IN.

Bekoff, M. 1972. The development of social interaction, play, and metacommunication in mammals: an ethological perspective. *Quart Rev Biol*: 412-434.

Bekoff, M., H. L. Hill and J. B. Mitton. 1975. Behavioral taxonomy in canids by discriminant function analysis. *Science* 90(4220):1223-1225.

Bekoff, M. and M. C. Wells. 1980. The social ecology and behavior of coyotes. *Adv Stud Anim Behav* 16: 251-338.

Bekoff, M., & Gese, E. M. 2003. Coyote (*Canis latrans*). *USDA National Wildlife Research Center-Staff Publications*, 224.

Bekoff, M., H. L. Hill and J. B. Mitton. 1975. Behavioral taxonomy in canids by discriminant function analysis. *Science* 190(4220): 1223-1225.

Bekoff, M., J. Diamond and J. B. Mitton. 1981. Life-history patterns and sociality in canids: body size, reproduction, and behavior. *Oecologia* 50 (3): 386-390.

Benecke, N. 1987. Studies on early dog remains from Northern Europe. *J Archaeol Sci* 14: 31-49.

Bentley, R. A., et al. 2009. *Handbook of Archaeological Theories*. Rowman & Littlefield, Lanham, Maryland.

Beregovoy, V. 2001. *The Hunting Laika Breeds of Russia*. Crystal Dreams Publ., Dover, TN, USA.

Berger, K. M. and E. M. Gese. 2007. Does interference competition with wolves limit the distribution and abundance of coyotes? *J Anim Ecol* 76: 1075-1085.

Berman, M. and I. Dunbar. 1983. The social behaviour of free-ranging suburban dogs. *Appl Anim Ethol* 10: 5-17. Berns, G. S., A. M. Brooks and M. Spivak. 2014. Scent of the familiar: an fMRI study of canine brain responses to familiar and unfamiliar human and dog odors. *Behav Process* doi:10.1016/j.beproc.2014.02.011.

Biknevicius, A. R. and C. B. Ruff. 1992. The structure of the mandibular corpus and its relationship to feeding behaviors in extant carnivores. *J Zool* 228: 479-507.

Biknevicius, A. R. and Van Valkenburgh, B. 1996. Design for killing: Craniodental adaptations of mammalian predators. Pp. 393-428 in J. L. Gittleman (ed.), *Carnivore Behavior, Ecology, and Evolution Vol. 2*.

Bininda-Emonds, O. R. P. 2000. Factors influencing phylogenetic inference: a case study using mammalian carnivores. *Mol Phylogenet Evol* 16(1): 113-126.

Björnerfeldt, S. 2007. Consequences of the Domestication of Man's Best Friend, The Dog. Universitatis Upsaliensis, Uppsala ISSN 1651-6214 ISBN 9789155468545.

Bock, W. J. 2004. Species: the concept, category and taxon. *J Zool Syst Evol Res* 42: 178-190.

Boitani, L. 1983. Wolf and dog competition in Italy. *Acta Zoologica Fennica* 174: 259-264.

Boitani, L., and P. Ciucci. 1995. Comparative social ecology of feral dogs and wolves. *Ethol Ecol Evol* 7: 49-72.

Boitani, L., P. Ciucci and A. Ortolani. 2007. Behaviour and social ecology of free-ranging dogs. Pp. 147-165 in P. Jensen (ed.), *The Behavioural Biology of Dogs*.

Boitani, L., et al. 1995. Population biology and ecology of feral dogs in Italy. Pp. 217-244 in J. Serpell, (ed.), *The Domestic Dog: Its evolution, behaviour and interactions with people*. Cambridge University Press, Cambridge

Bökönyi, S. 1989. Definitions of animal domestication. Pp. 24-27 in J. Clutton-Brock (ed.), *The Walking Larder. Patterns of Domestication, Pastoralism, and Predation*. Unwin Hyman, London.

Bonanni, R., et al. 2010. Effect of affiliative and agonistic relationships on leadership behavior in free-ranging dogs. *An Behav* 79: 981-991.

Bonanni, R., et al. 2011. Free-ranging dogs assess the quantity of opponents in intergroup conflicts. *Anim Cog* 14(1): 103-115.

Bonanni, R. and S. Cafazzo. 2014. The social organization of a population of free-ranging dogs in a suburban area of Rome: A reassessment of the effects of domestication on dogs' behaviour. Pp. 65-104 in J. Kaminski and S. Marshall-Pescini (eds.), *The Social Dog: behaviour and cognition*.

Bottema, S. 1989. Some observations on modern domestication processes. Pp. 31-45 in J. Clutton-Brock (ed.), *The Walking Larder. Patterns of Domestication, Pastoralism, and Predation*.

Boudadi-Maligne, M. and G. Escarguel. 2014. A biometric re-evaluation of recent claims for Early Upper Palaeolithic wolf domestication in Eurasia. *J Archaeo Sci* 45: 80-89.

Bowen, W. D. 1981. Variation in coyote social organization. The influence of prey size. *Can J Zool* 59: 639-652.

Boyd, D. K. and M. D. Jimenez. 1994. Successful rearing of young by wild wolves without mates. *J Mamm* 75(1): 14-17.

Boyko, A. R., et al. 2009. Complex population structure in African village dogs and its implications for inferring dog domestication history. *P Natl Acad Sci USA* 106: 13903-13908.

Boyko, A. et al. 2010. A simple genetic architecture underlies morphological variation in dogs. *PLOS* 8(8):e1000451 doi: 10.1371/journal.pbio.1000451

Bradshaw, J. 2011. *Dog Sense: How the New Science of Dog Behavior Can Make You a Better Friend to Your Pet*. Basic Books: Perseus

Bradshaw, J. W. S. and H. M. R. Nott. 1995. Social and communication behavior of companion dogs. Pp. 115-130 in J. Serpell (ed.), *The Domestic Dog: Its evolution, behaviour and interactions with people*.

Bradshaw, J. W., E. J. Blackwell and R. A. Casey. 2009. Dominance in domestic dogs—useful construct or bad habit? *J Veterinary Behav Clinical Appl and Rsch* 4(3): 135-144.

Brewer, D., T. Clark and A. Phillips. 2001. *Dogs in Antiquity: Anubis to Cerberus the Origins of the Domestic Dogs*. Aris & Phillips, Warminster, UK

Brisbin, I. L. Jr. et al. 1994. The New Guinea singing dog: taxonomy, captive studies and conservation priorities. *Sci New Guinea* 20: 27-38.

Brower, A. V. Z., R. DeSalle and A. Vogler. 1996. Gene trees, species trees, and systematics. *Annu Rev Ecol Syst* 27: 423-450.

Brown, S. K., et al. 2011. Phylogenetic distinctiveness of Middle Eastern and Southeast Asian village dog Y chromosomes illuminates dog origins. *PLoS One* 6(12): e28496 doi: 10.1371/journal.pone.0028496

Bromham, L. 2011. The genome as a life-history character: why rate of molecular evolution varies between mammal species. *Phil Trans Royal Soc B: Biol Sci* 366: 2503-2513.

Brothwell, D. R. and E. Higgs eds. 1969. *Science in Archaeology*. Tames and Hudson, London.

Brugal, J-P. and M. Boudadi-Maligne. 2011. Quaternary small to large canids in Europe: taxonomic status and biochronological contribution. *Quatern Int* 243: 171-182.

Burghardt, G. M. and J. L. Gittleman. 1990. Comparative behavior and phylogenetic analyses: New wine, old bottles. *Interpretation and explanation in the study of animal behavior* 2: 192-225.

Burleigh, R., et al. 1977. A further consideration of Neolithic dogs with a special reference to a skeleton from Grime's Graves (Norfolk), England. *J Archaeol Sci* 4: 353-366.

Butler, J. R. A. 1998. *The Ecology of Domestic Dogs, Canis familiaris, in the Communal Lands of Zimbabwe*. Ph.D. Thesis, Tropical Resource Ecology Programme, University of Zimbabwe.

Butler, J. R. A. and J. T. du Toit. 2002. Diet of free-ranging domestic dogs (*Canis familiaris*) in rural Zimbabwe: implications for wild scavengers on the periphery of wildlife reserves. *Anim Conserv* 5: 29-37.

Butler, J. R. A., J. T. Du Toit and J. Bingham. 2004. Free-ranging domestic dogs (*Canis familiaris*) as predators and prey in rural Zimbabwe: threats of competition and disease to large wild carnivores. *Biol Conserv* 115: 369-378.

Butzer, K. W. 1982. *Archaeology as Human Ecology: Method and Theory for a Contextual Approach*. Cambridge University Press.

Cafazzo, S., et al. 2010. Dominance in relation to age, sex, and competitive contexts in a group of free-ranging domestic dogs. *Behav Ecol* 21(3): 443-455.

Cafazzo S, et al. 2014. Social Variables Affecting Mate Preferences, Copulation and Reproductive Outcome in a Pack of Free-Ranging Dogs. *PLoS ONE* 9(6): e98594 doi:10.1371/journal.pone.0098594

Cahir, F. D. and I. Clark. 2013. The historic importance of the dingo in aboriginal society in Victoria (Australia): a reconsideration of the archival record. *Anthrozoös* 26(2): 185-198.

Calboli, F.C., et al. 2008. Population structure and inbreeding from pedigree analysis of purebred dogs. *Genetics* 179: 593-601.

Cannon, C. H., R. J. Morely and A. B. G. Bush. 2009. The current refugial rainforests of Sundaland are unrepresentative of their biogeographic past and highly vulnerable to disturbance *PNAS* 106 (27): 11188-11193.

Carbone, C. et al. 1999. Energetic constraints on the diet of terrestrial carnivores. *Nature* 402: 286-288.

Carbone, C., A. Teacher and J. M. Rowcliffe 2007. The costs of carnivory. *PLoS Biol* 5, e22. doi:10.1371/journal.pbio.0050022

Carpenter, C. A. 1963. Basenji-like dogs of Thailand and Borneo. *Sarawak Mus J* 11: 266-267.

Case, L. 2008. ASAS CENTENNIAL PAPER: Perspectives on domestication: The history of our relationship with man's best friend. *J Anim Sci* 86: 3245-3251.

Casinos, A., et al. 1986. On the allometry of long bones in dogs (*Canis familiaris*). *J Morphol* 190: 73-79.

Catling, P. C., L. K. Corbett and A. E. Newsome. 1992. Reproduction in captive and wild dingoes (*Canis familiaris dingo*) in temperate and arid environments of Australia. *Wildlife Res* 19(2): 195-209.

Cherin, M., et al. 2013. *Canis etruscus* (Canidae, Mammalia) and its role in the faunal assemblage from Pantalla (Perugia, central Italy): a comparison with the Late Villafranchian large carnivore guild of Italy. *Bollettino della Società Paleontologica Italiana* 52(1): 11-18.

Cherin, M., et al. 2014. Re-defining *Canis etruscus* (Canindae, Mammalia): a new look into the evolutionary history of early Pleistocene dogs resulting from the outstanding fossil record from Pantalla (Italy). *J Mammal Evol* 21: 95-110.

Chimpanzee Sequencing and Analysis Consortium. 2005. Initial sequence of the chimpanzee genome and comparison with the human genome. *Nature* 437: 69-87.

Christiansen, P. and J. S. Adolfssen. 2005. Bite forces, canine strength and skull allometry in carnivores (Mammmalia, Carnivora). *J Zool* 266: 133-151.

Christiansen, P. and S. Wroe. 2007. Bite forces and evolutionary adaptations to feeding ecology in carnivores. *Ecology* 88: 347-358.

Churchill, S. E. 1993. Weapon technology, prey size selection, and hunting methods in modern hunter-gatherers: implications for hunting in the Palaeolithic and Mesolithic. *Archeological Papers of the American Anthropological Association* 4: 11-24.

Clark, K. M. 2000. Dogged persistence: the phenomenon of canine skeletal uniformity in British prehistory. Pp. 163-169 in S. J. Crockford (ed.), *Dogs through Time: An Archaeological Perspective.*

Clark, P. U., et al. 2009. The Last Glacial Maximum. *Science* 325 (5941): 710-714.

Claridge, A. W. and R. Hunt. 2008. Evaluating the role of the Dingo as a trophic regulator: Additional practical suggestions. *Ecol Manage Restor* 9(2): 116-119.

Clifton, M. 2001. Street dogs keep the developing world from going to the rats. *Animal People* 10(6): 1/6.

Clifton, M. 2002. When the dogs are away, the monkeys will play. *Animal People* 11(1): 1/6-8.

Clutton-Brock, J. 1969. The origins of the dog. Pp. 303-309 in D. R. Brothwell and E. Higgs (eds.), *Science in Archaeology*. Thames and Hudson, London.

Clutton-Brock, J. 1981. *Domesticated Animals From Early Times*. British Museum (Natural History) & Heinemann, London.

Clutton-Brock, J. 1984. Dog. Pp. 198-211 in I. L. Mason (ed.), *Evolution of Domestic Animals*. Longman, London.

Clutton-Brock, J. 1992. The process of domestication. *Mammal Rev* 22: 79-85.

Clutton-Brock, J. 1995. Origins of the dog: domestication and early evolution. Pp. 7-20 in J. Serpell (ed.), *The Domestic Dog: Its Evolution, Behaviour and Interactions with People*.

Clutton-Brock, J. (ed.). 2014. *The Walking Larder*. Routledge, NY.

Clutton-Brock, J. and N. Noe-Nygaard. 1990. New osteological and C-isotope evidence on Mesolithic dogs: companions to hunters and fishers at Starr Carr, Seamr Carr and Kangemose. *J Archaeol Sci* 17: 643-653.

Cohen, J. 2007. Relative differences: the myth of 1%. *Science* 316: 1836.

Cohn, J. 1997. How wild wolves became domestic dogs: research sheds new light on the origin of humanity's most intimate quadruped ally. *BioScience* 47(1): 725-728.

Cole, B. F. and P. E. Koerper. 2002. The domestication of the dog in general—and dog burial research in the Southeastern United States. *J Alabama Acad Sci* 73(4): 174-179.

Colton, H. S. 1970. The aboriginal Southwestern Indian dog. *Am Antiq* 35(2): 153-159. Commission of the International Society for Zoological Nomenclature. 2003. Opinion 2027 (Case 3010). Usage of 17 specific names based on wild species which are pre-dated or contemporary with those based on domestic animals (Lepidoptera, Osteichtyes, Mammalia): conserved. *Bull Zool Nomenclature* 60(1): 81-84.

Connor, J. L. 1975. Genetic mechanisms controlling the domestication of a wild house mouse population (*Mus musculus* L.). *J Comp Physiol Psychol* 89(2): 118-130.

Conroy, C. J. and M. vanTuinen. 2003. Extracting time from phylogenies: positive interplay between fossil and genetic data. *J Mammal* 84(2): 444-455.

Coppinger, R. A. and C. K. Smith. 1983. The domestication of evolution. *Environ Conserv* 10(4): 283-292.

Coppinger, R. and L. Coppinger. 2001. *Dogs: A Startling New Understanding of Canine Origin, Behavior and Evolution*. Scribner, NY.

Corbett, L. K. 1988. Social dynamics of a captive dingo pack: population regulation by dominant female infanticide. *Ethology* 78(3): 177-198.

Corbett, L. K. 1995. *The Dingo in Australia and Asia*. Cornell Univ. Press, Ithaca, NY.

Corbett, L. K. 2004. Dingo, *Canis lupus dingo* (Meyer, 1793). Pp. 223-230 in C. Sillero-Zubiri, M. Hoffmann and D. Macdonald (eds.) *Canids: foxes, wolves, jackals and dogs*. IUCN, Cambridge, UK.

Corbett, L. K. and A. Newsome. 1987. The feeding ecology of the dingo, III. Dietary relationships with widely fluctuating prey populations in arid Australia: an hypothesis of alternation of predation. *Oecologia* 74: 215-227.

Cox, M. P. 2008. Accuracy of molecular dating with the rho statistic: deviations from coalescent expectations under a range of demographic models. *Hum Biol* 80(4): 335-357.

Coyne, J. A. and H. A. Orr. 2004. *Speciation*. Sinauer Assoc., Inc., Sunderland, MA.

Cranbrook, E. 1988. The contribution of archaeology to the zoogeography of Borneo, with the first record of a wild canid of early Holocene age. *Fieldiana Zool Ser* 42: 1-7.

Crandall, K. A., et al. 2000. Considering evolutionary processes in conservation biology. *TREE* 15(7): 290-295.

Crisler, L. 1958. *Artic Wild*. Harper and Row, NY.

Crisler, L. 1968. *Captive Wild*. Harper and Row, NY.

Crockford, S. J. (ed.). 2000. *Dogs through Time: An Archaeological Perspective; Proceedings of the 1st ICAZ Symposium on the History of the Domestic Dog; Eighth Congress of the International Council for Archaeozoology (ICAZ98), August 23-29, 1998, Victoria, BC, Canada.* BAR International Series 889, Archaeopress, Oxford.

Crockford, S. J. and Y. V. Kuzmin. 2012. Comments on Germonpré et al., *Journal of Archaeological Science* 36, 2009 "Fossil dogs and wolves from Palaeolithic sites in Belgium, the Ukraine and Russia: osteometry, ancient DNA and stable isotopes", and Germonpré, Lázkičková-Galetová, and Sablin, Journal of Archaeological Science 39, 2012 "Palaeolithic dog skulls at the Gravettian Předmostí site, the Czech Republic". *J Archaeol Sci* 39(8): 2797 -2801.

Cronin, M. A. 1993. Mitochondrial DNA in wildlife taxonomy and conservation biology: cautionary notes. *Wildlife Soc B* 21: 339-348.

Cronin, M. A., et al. 1991. Interspecific and intraspecific mitochondrial DNA variation in North American bears (*Ursus*). *Can J Zool* 69(12): 2985-2992.

Crowther, M. S., et al. 2014. An updated description of the Australian dingo (*Canis dingo* Meyer, 1793). *J Zool* 293(3): 192-203.

Curtis, A. and B. Van Valkenburgh. 2014. Beyond the sniffer: frontal sinuses in Carnivora. *Anat Rec* 297: 2047-2064.

Danchin, É. et al. 2011. Beyond DNA: integrating inclusive inheritance into an extended theory of evolution. *Nat Rev Genet* 12: 475-486.

Daniels, T. J. 1983a. The social organization of free-ranging urban dogs. I. Non-estrous social behavior. *Appl Anim Ethol* 10: 341-363.

Daniels, T. J. 1983b. The social organization of free-ranging urban dogs. II. Estrous groups and the mating system. *Appl Anim Ethol* 10: 365-373.

Daniels, T. J. and M. Bekoff. 1989a. Population and social biology of free-ranging dogs *Canis familiaris*. *J Mammal* 70(4): 754-762.

Daniels, T. J. and M. Bekoff. 1989b. Feralization: the making of wild domestic animals. *Behav Process* 19(1): 79-94.

Darmesteter, J. (ed.) 1880. Fargard XIII. The Dog. Pp. 151-171 in *The Zend-Avesta, part I: The Vendîdâd*. Clarendon Press, Oxford.

Das, G. 2009. Personal messages re: Indian village dog coprophagy.

Dávalos, L. M. and A. L. Russell. 2014. Sex-biased dispersal produces high error rates in mitochondrial distance-based and tree-based species delimitation. *J Mamm* 95(4): 781-791.

Davis, S. and F. R. Valla. 1978. Evidence for domestication of the dog 12,000 years ago in the Natufian of Israel. *Nature* 276: 608-610.

Dayan, T. 1994. Early domestic dogs of the Near East. *J Archaeol Sci* 21: 633-640.

Dayan, T. and E. Galili. 2001. A preliminary look at some new domesticated dogs from submerged Neolithic sites off the Carmel Coast. Pp. 29-30 in S. J. Crockford (ed.), *Dogs through Time: An Archaeological Perspective.*

Dayan, T., D. Wool and D. Simberloff. 2002. Variation and covariation of skulls and teeth: modern carnivores and the interpretation of fossil mammals. *Paleobiology* 28(4): 508-526.

Deguilloux, M. F., et al. 2009. Ancient DNA supports lineage replacement in European dog gene pool: insight into Neolithic southeast France. *J Archaeol Sci* 36 (2): 513-519.

De Queiroz, K. 2005. A unified concept of species and its consequences for the future of taxonomy. *P Cal Acad Sci 56, Supplement* 1(18): 196-215.

Derr, M. 2011. *How the Dog Became the Dog: from Wolves to Our Best Friends.* Overlook, Duckworth, NY.

Dinets, V. 2007. The history of dog domestication. http://dinets.travel.ru/dogs.htm. Downloaded 7/8/2007.

Dinets, V. 2015. The *Canis* tangle: a systematics overview and taxonomic recommendations. *Vavilov Journal of Genetics and Breeding* 19: 286-291. Ding, Z-L., et al. 2012. Origins of domestic dog in Southern East Asia is supported by analysis of Y-chromosome DNA. *Heredity* 108: 507-514.

Doebeli M. and U. Dieckmann. 2002. Speciation along environmental gradients. *Nature* 421: 259-264.

Dorey, N. R., M. A. Udell and C. D. Wynne. 2009. Breed differences in dogs' sensitivity to human points: a meta-analysis. *Behav Processes* 81(3): 409-415.

Downs, J. F. 1960. Domestication: an examination of the changing social relationships between man and animals. *Kroeber Anthropol Soc Papers 22*. U. C. Berkeley, CA.

Drake, A. G. 2011. Dispelling the dogma: an investigation of heterochrony in dogs using 3D geometric morphometric analysis of skull shape. *Evol Dev* 13(2): 204-213.

Drake, A. G. and C. P. Klingenberg. 2010. Large-scale diversification of skull shape in domestic dogs: disparity and modularity. *Am Nat* 175: 289-301.

Drake, A. G., M. Coquerelle and G. Colombeau. 2015. 3D morphometric analysis of fossil canid skulls contradicts the suggested domestication of dogs during the late Paleolithic. *Scientific Reports* Volume: 5: Article number 8299, DOI: doi: 10.1038/srep08299

Drews, C. 1993. The Concept and Definition of Dominance in Animal Behaviour. *Behaviour* 125 (No. 3/4): 283-313.

Driscoll, C. A., D. W. Macdonald and S. J. O'Brien. 2009. From wild animals to domestic pets, an evolutionary view of domestication. *PNAS* 106(1): 9971-9978.

Driscoll, C. A. and D. W. Macdonald. 2010. Top dogs: wolf domestication and wealth. *J Biol* 9:10, http://jbiol.com/content/9/2/10

Druzkhova, A. S., et al. 2013. Ancient DNA analysis affirms the canid from Altai as a primitive dog. *PLoS One* 8(3):e57754. doi:10.1371/journal.pone.0057754

Duboule, D. 1995. Vertebrate Hox genes and proliferation: an alternative pathway to homeosis? *Curr Opin Gen & Devel* 5(4): 525-528.

Dung, V. V., et al. 1993. A new species of living bovid from Vietnam. *Nature* 363: 443-445.

Dupré, J. 2006. Scientific classification. *Theory Culture Society*: 23(2/3): 30.

Edgaonkar, A. and R. Chellam. 2002. Food habit of the leopard, *Panthera pardus*, in the Sanjay Gandhi National Park, Maharashtra, India. *Mammalia* 66(3): 353-360.

Epstein, H. 1971. *The Origin of the Domestic Animals of Africa*. Volume I. Africana Pub. Corp, NY.

Eswaran, V., H. Harpending and A. R. Rogers. 2005. Genomics refutes an exclusively African origin of humans. *J Hum Evol* 49: 1-18.

Evin, A., et al. 2013. The long and winding road: identifying pig domestication through molar size and shape. *J Archaeo Sci* 40(1): 735-743.

Ewer, R. F. 2013. *Ethology of mammals*. Springer, NY.

Feddersen-Petersen, D. 1986. Observations of social play in some species of Canidae. *Zool Anz* 217(1/2): 130-144.

Feddersen-Petersen, D. 2007. Social behaviour of dogs and related canids. Pp. 105-119 in P. Jensen (ed.), *The Behavioural Biology of Dogs*.

Felician, P. 2012. SIRT1 and anxiety. *Nat Genet* 44 (120):doi:10.1038/ng.1093

Felsenstein, J. 1985. Phylogenies and the comparative method. *Am Nat* 125: 1-15.

Fentress, J. C. 1967. Observations on the behavioral development of a hand-reared male timber wolf. *Am Zool* 7: 339-351.

Ferguson, W. W. 1981. The systematic position of *Canis aureus lupaster* (Carnivora: Canidae) and occurrence of *Canis lupus* in North Africa, Egypt and Sinai. *Mammalia* 45: 459-465.

Fiennes, R. and A. Fiennes. 1968. *The Natural History of Dogs*. Bonanza Books, NY. Finlayson, C. 2009. *The Humans Who Went Extinct*. Oxford Univ. Press, NY. Fondon, J. W. and H. R. Garner. 2004. Morphological origins of rapid and continuous morphological evolution. *Proc Natl Acad Sci USA* 101: 18058-18063.

Font, E. 1987. Spacing and social organization: Urban stray dogs revisited. *Appl Anim Behav Sci* 17 (3-4): 319-328.

Foote, M. 2001. Inferring temporal patterns of preservation, origination, and extinction from taxonomic survivorship analysis. *Paleobiology* 27(4): 602-630.

Forster, P. and S. Matsumura. 2005. Did early humans go north or south? *Science* 308: 965-966.

Fox, M. W. 1969. Ontogeny of prey-killing behavior in Canidae. *Behaviour* 35(3): 259-272.

Fox, M. W. 1969. The anatomy of aggression and its ritualization in Canidae: a developmental and comparative study. *Behaviour* 35: 242-258.

Fox, M. W. 1971. *Behaviour of Wolves, Dogs and Related Canids*. Dogwise Publishing, Wenatchee, WA.

Fox, M. W. 1972. Patterns and problems of socialization in hand-reared wild canids: an evolutionary and ecological perspective. *Z Tierpsychol* 31(3): 281-288.

Fox, M. W. 1973. Origin of the dog and effects of domestication. *AKC Gazette* 90(1): 33-35.

Fox, M. W. 1978. *The Dog: its domestication and behavior*. Garland Publishing Inc., NY.

Fox, M. W., A. M. Beck and E. Blackman. 1975. Behavior and ecology of a small group of urban dogs (*Canis familiaris*). *Appl Anim Ethol* 1: 119-137.

Frank, H. 2011. Wolves, dogs, rearing and reinforcement: complex interactions underlying species differences in training and problem-solving performance. *Behav Genet* 41(6): 830-839.

Frank, H. and M. G. Frank. 1982. On the effect of domestication on canine social development and behavior. *Appl Anim Ethol* 8: 507-525.

Frank, H. and M. G. Frank. 1983. Inhibition training in wolves and dogs. *Behav Process* 8: 363-377.

Frank, H. and M. G. Frank. 1987. The University of Michigan canine information-processing project (1979-1981). Pp. 143-167 in H. Frank (ed.), *Man and Wolf*. Dordrecht: Dr. W. Junk.

Frank, H. and M. G. Frank. 1985. Comparative manipulation-test performance in ten-week-old wolves (*Canis lupus*) and Alaskan malamutes (*Canis familiaris*): a Paigetian interpretation. *J. Comp. Psych.* 99(3): 266-274.

Freedman, A. H., et al. 2014. Genome sequencing highlights the dynamic early history of dogs. *PLoS gen* 10(1): e1004016

Freedman, D. G., J. A. King and O. Elliot. 1961. Critical periods in the social development of dogs. *Science* 133: 1016-1017. Fuller, T. K., et al. 1989. The ecology of three sympatric jackal species in the Rift Valley of Kenya. *Afr J Ecol* 27: 313-323.

Gácsi, M., et al. 2005. Species-specific differences and similarities in the behavior of hand-raised dog and wolf pups in social situations with humans. *Dev Psychobiol* 47: 111-122.

Gácsi, M., et al. 2009a. Effects of selection for cooperation and attention in dogs. *Behav Brain Funct* 5(1), 31. doi:10.1186/1744-9081-5-31

Gácsi, M., et al. 2009b. Explaining dog wolf differences in utilizing human pointing gestures: selection for synergistic shifts in the development of some social skills. *PLoS ONE* 4(8): e6584. doi:10.1371/journal.pone.0006584

Gallant, J. 2002. *The Story of the African Dog*. Univ. of Natal Press, Pietermaritzburg, ZA.

Galton, F. 1883. *Inquiry into human faculty and its development*. London. Macmillan.

García, N. and E. Vírgós. 2007. Evolution of community composition in several carnivore palaeoguilds from the European Pleistocene: the role of interspecific competition. *Lethia* 40: 33-44.

Gardner, A. L. and V. Hayssen. 2004. A guide to constructing and understanding synonymies. American Society of Mammalogists, *Mammalian Species* No. 739: 1-17.

Garrido, G. and A. Arribas. 2008. *Canis accitanus* nov. sp., a new small dog (Canidae, Carnivora, Mammalia) from the Fonelas P-1 Plio-Pleistocene site (Gaudix basin, Granada, Spain). *Geobios* 41: 751-761. 224 Dawn of the Dog

Geffen E., et al. 1996. Size, life-history traits, and social organization in the Canidae: A reevaluation. *Am Nat* 147: 140-160.

Gehring, W. J. 1998. *Master Control Genes in Development and Evolution: The Homeobox Story*. Yale University Press.

Geist, V. 1991. Taxonomy: on an objective definition of subspecies, taxa as legal entities, and its application to Rangifer Tarandus Lin. 1758. Pp. 1-36 in C. Butler and S. P. Mahoney (eds.), *Proceedings 4th North American Caribou Workshop St. John's Newfoundland October 31-November 3, 1989*; Newfoundland and Labrador Wildlife Division.

Gentry, A., J. Clutton-Brock and C. P. Groves. 2004. The naming of wild animal species and their domestic derivatives. *J Archaeol Sci* 31: 645-651.

Gerbault, P., et al. 2014. Storytelling and story testing in domestication. *PNAS* doi:10.1073/pnas.1400425111

Germonpré, M., *et al.* 2009. Fossil dogs and wolves from Paleolithic sites in Belgium, the Ukraine and Russia: Osteometry, ancient DNA and stable isotopes. *J Archaeol Sci* 36: 473-490.

Germonpré, M., M. Laznickova-Galetova and M. V. Sablin. 2012. Palaeolithic dog skulls at the Gravettian Predmosti site, the Czech Republic. *J Archaeol Sci* 39: 184-202.

Germonpré, M., et al. 2013. Palaeolithic dogs and the early domestication of the wolf: a reply to the comments of Crockford and Kuzmin (2012). *J Archaeol Sci* 40: 786-792.

Germonpré, M., et al. 2015. Palaeolithic dogs and Pleistocene wolves revisited: a reply to Morey (2014) *J Archaeol Sci* 54: 210-216.

Ghosh, B., D. K. Choudhuri and B. Pal. 1984. Some aspects of the sexual behaviour of stray dogs, *Canis familiaris*. *Appl Anim Beh Sci* 13: 113-127.

Gifford-Gonzalez, D. 1991. Bones are not enough: analogues, knowledge, and interpretive strategies in zooarchaeology. *J Anthropo Archaeo* 10(3): 215-254.

Gittleman, J. L. 1985. Carnivore body size: ecological and taxonomic correlates. *Oecologia* 67: 540-554.

Gittleman, J. L. 2013. *Carnivore Behavior, Ecology, and Evolution. Vols. 1 & 2*. Springer Science & Business Media, Berlin.

Gingeras, T. R. 2007. Origin of phenotypes: genes and transcripts. *Genome Res* 17: 682-690.

Ginsberg, B. E. 1968. Genotypic factors in the ontogeny of behavior. Pp. 12-17 in Jules H. Masserman, MD (ed.), *Science and Psychoanalysis. Vol XII: Animal and Human*. Grune & Stratton, NY.

Giorgi, E. 2015. From Many, One: genetic chimerism. http://www.the-scientist.com//?articles.view/articleNo/42476/title/From-Many--One/

Glen, A. S., et al. 2007. Evaluating the role of the dingo as a trophic regulator in Australian ecosystems. *Aust Ecol* 32: 492-501.Gollan, K. 1982. *The prehistoric dingo*. Ph.D. diss. Canberra: Australian National University.

Gompper, M. E. 2013. Chapter 1. The dog-human-wildlife interface: assessing the scope of the problem. Pp. -55 in M. E. Gompper (ed.), *Free-ranging dogs and wildlife conservation*. Oxford University Press.

Gonzalez, T. 2012. *The pariah case: some comments on the origin and evolution of primitive dogs and on the taxonomy of related species*. PhD Thesis. School of Archaeology and Anthropology, Australian National University, Canberra.

Goodwin, D., J. W. S. Bradshaw and S. M. Wickens. 1997. Paedomorphosis affects agonistic visual signals of domestic dogs. *Anim Behav* 53: 297-304.

Grant, P. R. and B. R. Grant. 2002. Adaptive radiation of Darwin's finches: Recent data help explain how this famous group of Galapagos birds evolved, although gaps in our understanding remain. *Am Scientist* 90(2): 130-139.

Grant, R. and P. Grant. 2010. Inaugural Article: Songs of Darwin's finches diverge when a new species enters the community. *PNAS* 107:20156-20163

Grant, P. R., R. Grant and K. Pentren. 2005. Hybridization in the recent past. *Am Nat* 16(1): 56 -67.

Graves, W. N. 2007. *Wolves in Russia: Anxiety through the Ages*. Detselig Ent. Ltd., Calgary, Alberta, Canada.

Gray, M. M., et al. 2009. Linkage disequilibrium and demographic history of wild and domestic canids. *Genetics* 181(4): 1493-1505.

Gray M. M., et al. 2010.The IGF1 small dog haplotype is derived from Middle Eastern grey wolves. *BMC Biol* 8:16. doi:10.1186/1741-7007-8-16

Grayson, D. K. (ed.). 2014. *Quantitative Zooarchaeology: Topics in the Analysis of Archaelogical Faunas*. Academic Press, Orlando.

Griffiths, P. E. and K. Stotz. 2006. Genes in the postgenomic era. *Theor Med Bioethics* 27(6): 499-521.

Grimm, D. 2015. Dawn of the dog. *Science* 348(6232): 274-279.

Haag, W. G. 1948. An osteometric analysis of some aboriginal dogs. *Univ. of Kentucky, Dept. Anthro Reports in Anthropology* 7(3): 107-264.

Habib, B., S. Shrotriya, S. and Y. V. Jhala, 2013. *Ecology and Conservation of Himalayan Wolf*. Wildlife Institute of India-Technical Report No. TR-2013/01, 46 pp.

Hall, S. S. 2012. Journey to the genetic interior: what was once known as junk DNA turns out to hold hidden treasures, says computational biologist Ewan Birney. *Sci Am* 307(4): 80-84.

Hallett, M. 1987. Two wolves, one land: could two closely related social predators share the same environment? *Terra* 25(6): 11-17.

Hamilton, S. 2009. Personal messages re: wolf coprophagy.

Hanebuth, T. J. J. and K. Stattegger. 2010. The stratigraphic evolution of the Sunda shelf during the past fifty thousand years. *Tropical Deltas of Southeast Asia - Sedimentology, Stratigraphy, and Petroleum Geology* 76: 189-200.

Harcourt, R. A. 1974. The dog in prehistoric and early historic Britain. *J Archaeol Sci* 1(2): 151-175.

Hare, B. and V. Woods. 2013. *The Genius of Dogs*. Dutton, NY.

Hare, B., et al. 2002. The domestication of social cognition in dogs. *Science* 298(5598): 1634-1636.

Hare, B., et al. 2005. Social cognitive evolution in captive foxes is a correlated by-product of experimental domestication. *Curr Biol* 15(3): 226-230.

Hare, B., et al. 2010. The domestication hypothesis for dogs' skills with human communication: a response to Udell et al. (2008) and Wynne et al. (2008). *Anim Behav* 79:e1-e6.

Harrison, D. L. 1973. Some comparative features of the skulls of wolves (*Canis lupus* Linn.) and pariah dogs (*Canis familiaris* Linn.) from the Arabian Peninsula and neighboring lands. *Bonn Zool Beitr* Heft 3 24/1973: 185-191.

Harrison, D. L. and P. J. Bates. 1991. *The Mammals of Arabia*. Harrison Institute, Kent, UK.

Hart, D. and R. W. Sussman. 2011. The influence of predation on primate and early human evolution: impetus for cooperation. Pp. 9-40 in L. Barrett (ed.), *Developments in Primatology: prognosis and prospects. Vol. 36*. Springer, NY.

Hartl, D. L. and Clark, A. 1997. *Principles of Population Genetics*. Sinauer Associates, Inc., Sunderland, MA.

Hasse, E. 2000. Comparison of reproductive biological parameters in male wolves and domestic dogs. *Z Säugetierkd* 65: 257-270.

Hawks, J. 2013. Significance of Neandertal and Denisovan genomes in human evolution. *Ann Rev Anthro* 42: 433-449.

He, Y., et al. 2010. Heteroplasmic mitochondrial DNA mutations in normal and tumour cells. *Nature* 464: 610-614.

Heaney, L. R. 1986. Biogeography of mammals in SE Asia: estimates of rates of colonization, extinction and speciation. *Biol J Linn Soc* 28: 127-165.

Hecht, J. and A. Horowitz. 2015. Introduction to dog behavior. Pp. 5-30 in E. Weiss, H. Mohan-Gibbons and S. Zawistowski (eds.), *Animal Behavior for Shelter Veterinarians and Staff*. John Wiley & Sons, Inc.

Hedrick, P. and L. Waits. 2005. What ancient DNA tells us. *Heredity* 94: 463-464.

Hemmer, V.-H. 1975. Zur Abstammung des Haushundes und zur Veränderung der relativen Hirngröße bei der Domestikation. *Zoologische Beiträge* 21/1: 97-104. [English Summary]

Hemmer, Helmut. 1976. Man's strategy in domestication—a synthesis of new research trends. *Experientia* 32(5): 663-666.

Hemmer, H. 1990. *Domestication: The Decline of Environmental Appreciation.* Cambridge Univ. Press, NY.

Herre, Wolf. 1970. The science and history of domestic animals. Pp. 257-272 in D. Brothwell and E. Higgs (eds.), *Science in Archaeology.* Praeger, NY.

Hetts, S. 2014. *12 Terrible Dog Training Mistakes That Owners Make That Ruin Their Dog's Behavior . . . And How To Avoid Them.* Animal Behavior Associates, Inc., Littleton, CO.

Hey, J. 2001. *Genes, Categories, and Species: The Evolutionary and Cognitive Causes of the Species Problem.* Oxford Univ. Press.

Higham, C. F. W., et al. 1980. An analysis of prehistoric canid remains from Thailand. *J Archaeol Sci* 7: 149-165.

Hillis, D. M. 1987. Molecular versus morphological approaches to systematics. *Ann Rev Ecol Syst* 18: 23-42.

Ho, S. Y., et al. 2008. The effect of inappropriate calibration: three case studies in molecular ecology. *PLoS One* 3(2), e1615

Ho, S. Y., et al. 2011. Time-dependent rates of molecular evolution. *Mol Ecol* 20(15): 3087-3101.

Hoelzer, G. A. 1997. Inferring phylogenies from mtDNA variation: mitochondrial gene-trees versus nuclear gene-trees revisited. *Evolution* 51: 622-626. Holden, C. 2006. An evolutionary squeeze on brain size. *Science* 5782: 1867.

Holliday, J. A. and S. J. Steppan. 2004. Evolution of hypercarnivory: the effect of specialization on morphological and taxonomic diversity. *Paleobiol* 30(1): 108-128.

Honeycutt, R. L. 2010. Unraveling the mysteries of dog evolution. *BMC Biol* 8: 20 doi: 10.1186/1741-7007-8-20.

Houghton, P. 1996. *People of the Great Ocean: Aspects of Human Biology of the Early Pacific.* Cambridge Univ. Press, NY.

Horard-Herbin, M-P., A. Tresset and J-D. Vigne. 2014. Domestication and uses of the dog in western Europe from the Paleolithic to the Iron Age. *Animal Frontiers* 4(3): 23-31.

Horowitz, A. (ed.) 2014. *Domestic Dog Cognition and Behavior.* Springer, Berlin.

Hsu, Y. and J. A. Serpell. 2003. Development and validation of a questionnaire for measuring behavior and temperament traits in pet dogs. *J Am Veterinary Med Assn* 223(9): 1293-1300.

Hubbard, C. L. B. and S. Lampson. 1966. *The Observer's Book of Dogs.* F. Warne Publ., NY.

Hughes, J. and D. W. Macdonald. 2013. A review of the interactions between free-roaming domestic dogs and wildlife. *Biol Conserv* 157: 341-351.

Hwang, D. G. and P. Green. 2004. Bayesian Markov chain Monte Carlo sequence analysis reveals varying neutral substitution patterns in mammalian evolution. *PNAS* 101(39): 13994-14001.

Ichhpujani, R. L., et al. 2008. Epidemiology of animal bites and rabies cases in India: a multicentric study. *J Commun Dis* 40(1): 27-36.

ICZN. 2003. Opinion 2027 (Case 3010). Usage of 17 specific names based on wild species which are pre-dated by or contemporary with those based on domestic animals (Lepidoptera,

Osteicthyes, Mammalia): conserved. *Bull Zool Nomenclature* 60(1): 81-84.

Irion, D. N., et al. 2005. Genetic variation analysis of the Bali street dog using microsatellites. *BMC Genet* 6(1):6 doi: 10.1186/1471-2156-6-6

Ivanoff, D. V. 2001. Partitions in the carnivoran auditory bulla: their formation and significance. *Mammal Rev* 31(1): 1-16.

James, H. V. A. and M. D. Petraglia. 2005. Modern human origins and the evolution of behavior in the later Pleistocene record of South Asia. *Curr Anthropol* 46(S5): S3-S27.

Janssens, L., R. Miller and S. Van Dongen, 2016. The morphology of the mandibular coronoid process does not indicate that Canis lupus chanco is the progenitor to dogs. Zoomorph: DOI 10.1007/s00435-015-0298-z

Jensen, P. (ed.). 2007. *The Behavioural Biology of Dogs*. Oxford: CABI International.

Jinam, T. A., et al. 2012. Evolutionary history of continental Southeast Asians: "Early Train" hypothesis based on genetic analysis of mitochondrial and autosomal DNA data. *Mol Biol Evol* 29 (11): 3513-3527.

Jiao, T. 2009. Correspondence about early Chinese dogs.

Johannes, J. 2003. The Basenji annual estrus: African origins. *The Basenji* 38(10): 10-11.

Johnson, C. N. and S. Wroe. 2003. Causes of extinction of vertebrates during the Holocene of mainland Australia: arrival of the dingo, or human impact? *The Holocene* 13(6): 941-948.

Joordens, J. C. A., et al. 2015. *Homo erectus* at Trinil on Java used shells for tool production and engraving. *Nature* doi: 10.1038/nature13962

Kahlke, R-D., et al. 2011. Western Palaearctic palaeoenvironmental conditions during the Early and early Middle Pleistocene inferred from large mammal communities, and implications for hominin dispersal in Europe. *Quaternary Sci Rev* 30 (11): 1368-1395.

Kaminski, J. and S. Marshall-Pescini (eds.). 2014. *The Social Dog: Behavior and Cognition*. Academic Press, San Diego, CA.

Kaulfuß, P. and D. S. Mills. 2008. Neophilia in domestic dogs (*Canis familiaris*) and its implication for studies of dog cognition. *Anim Cogn* 11: 553-556.

Keele, C., J. Asteinza and E. Fromm. 1964. Psychosomatics of fear in foxes. *Georgia Acad Sci* 22: 64-69.

Kershenbaum A. and Garland E. 2015. Quantifying animal sequence similarity: which metric performs best? *Methods in Ecology and Evolution* 6:1452-1461.

Kershenbaum, A. 2015. The Cooperative Carnivore Vocalization Project, personal communication.

Kershenbaum, A., et al. 2016. Disentangling canid howls across multiple species and subspecies: Structure in a complex communication channel. *Behav Processes* 124: 149-157.

Khanna, C., et al. 2006. The dog as a cancer model. *Nature Biotechnology* 24(9): 1065-1066.

Kieser, J. A. and H. T. Groenveld. 1992. Comparative morphology of the mandibulodental complex in wild and domestic canids. *J Anat* 180: 419-424.

Kimball, W. H. and L. B. Matrin (eds.). 1993. *Species, Species Concepts, and Primate Evolution*. Plenum Press, NY.

Kirsch, J. A.W. and J. I. Johnson. 1983. Phylogeny though brain traits: trees generated by neural characters. *Brain Behav Evolut* 22: 60-69.

Kiyu, A. 1981. The dogs and their possible influence on health of the longhouse people of the seventh division of Sarawak. *Sarawak Museum J* 50: 97-100.

Klinghammer, E. and P. A. Goodman. 1987. *The Management and Socialization of Captive Wolves (Canis lupus) at Wolf Park. Ethology Series No. 2*. North American Wildlife Park Foundation, Inc., Battle Ground, IN.

Klinghammer, E. and P. A. Goodmann. 1987. Socialization and management of wolves in captivity. Pp. 31-60 in H. Frank (ed.), *Man and wolf: Advances, issues and problems in captive wolf research*. Dordrecht: W. Junk Publishers.

Knowles, L.L. and B. C. Carstens, 2007. Delimiting species without monophyletic gene trees. *System Biol* 56(6):.887-895.

Koler-Matznick, J. 2002. The origin if the dog revisited. *Anthrozöos* 15: 98-118.

Koler-Matznick, J., I. l. Brisbin, Jr. and M. Feinstein. 2001. *An ethogram of the New Guinea singing dog, Canis hallstromi*. New Guinea Singing Dog Conservation Society, Central Point, OR.

Koler-Matznick, J., et al. 2003. An updated description of the New Guinea singing dog (*Canis hallstromi*, Troughton 1957). *J Zool London* 261(2): 109-118.

Koler-Matznick, J. and M. Stinner. 2011. First report of captive New Guinea dingo *(Canis dingo hallstromi)* den-digging and parental behavior. *Zoo Biol* 30: 445-450.

Koler-Matznick, J. and B. C. Yates. 2015. The craniodental adaptations of the generalized dog, *Canis familiaris*. MS in preparation available from J. Koler-Matznick.

Kolig, E. 1973. Aboriginal man's best foe? *Mankind* 9: 122-123.

Kolig, E. 1978. Aboriginal dogmatics: canines in theory, myth and dogma. *Bijdragen tot de Taal-, Land-en Volkenkunde, Deel 134, 1ste Afl., Anthropologica XX*: 84-115.

Koopmans, M., et al. 2005. Chimerism in kidneys, livers and hearts of normal women: Implications for transplantation studies. *Am J Transplant* 5: 1495-1502.

Kopaliani, N., et al. 2014. Gene flow between wolf and shepherd dog populations in Georgia (Caucasus). *J. Heredity* 105(3): 345-353.

Korpimäki, E. and C. J. Krebs. 1996. Predation and population cycles of small mammals. *BioSci* 46(10): 754-764.

Kruska, D. 1988. Mammalian domestication and its effect on brain structure and behavior. Pp. 211-250 in H. J. Jerison and I. Jerison (eds.), *Intelligence and Evolutionary Biology*. Springer-Verlag, Berlin.

Kubinyi, E., Z. Virányi and Á. Miklósi. 2007. Comparative social cognition: from wolf and dog to humans. *Comp Cogn Behav Rev* 2: 26-46.

Kubinyi, E., P. Pongrácz and Á. Miklósi 2009. Dog as a model for studying conspecific and heterospecific social learning. *J Veterinary Behav: Clinical Appli Resch* 4(1): 31-41.

Kukekova, A. V., et al. 2011. Mapping loci for fox domestication: deconstruction/reconstruction of a behavioral phenotype. *Behav Genet* 41: 593-606.

Kukekova, A. V., et al. 2012. Genetics of behavior in the silver fox. *Mamm Genome* 23 (1-2): 164-177.

Kurtén, B. 1968. *Pleistocene Mammals of Europe*. Aldine Pub Co., Chicago.

Künzl, C., et al. 2003. Is a wild mammal kept and reared in captivity still a wild animal? *Hormones and Behavior* 43: 187-196.

Laidlaw, J., et al. 2007. Elevated basal slippage mutation rates among the Canidae. *J Hered* 98(5): 452-460.

Larkin, J. C. 2009. Morphological evolution: by any means necessary? *Curr Biol* 19 (20): R953-R954.

Larson, G., et al. 2012. Rethinking dog domestication by integrating genetics, archeology, and biogeography. *PNAS* 109: 8878-8883.

Larson, G. and D. G. Bradley. 2014. How much is that in dog years? The advent of canine population genomics. *PloS Genet* 10(1): e1004093.doi:10.1371/journal.pgen.1004093

Larson, G., et al. 2014. Current perspectives and the future of domestication studies. *PNAS USA* 111(17): 6139-6146.

Lau, A. N., et al. 2008. Horse domestication and conservation genetics of Przewalski's horse inferred from sex chromosomal and autosomal sequences. *Mol Biol Evol* 26(1): 199-208.

Lawrence, B. 1967. Early domestic dogs. *Zeitschrift Säugetierk* 32(1): 44-59.

Lawrence, B. 1968. Antiquity of large dogs in North America. *Tebiwa* 11(2): 43-49.

Lawrence, B. and W. H. Bossert. 1967. Multiple character analysis of *Canis lupus, latrans,* and *familiaris* with a discussion of the relationships of *Canis niger. Am Zool* 7: 223-232.

Lawrence, B. and W. H. Bossert. 1969. The cranial evidence for hybridization in New England *Canis. Brevoria* 330: 1-13.

Leach, M. 1961. *God had a Dog: Folklore of the Dog.* Rutgers University Press.

Lee, R. B. and R. H. Daly (eds.). 1999. *The Cambridge Encyclopedia of Hunters and Gatherers.* Cambridge University Press, MA.

Lehner, P. N. 1998. *Handbook of ethological methods.* Cambridge University Press.

Leonard, J. A., et al. 2002. Ancient DNA evidence for Old World origin of New World dogs. *Science* 298: 1613-1616.

Leonard, J. A., et. al. 2007. Megafaunal extinctions and the disappearance of a specialized wolf ecomorph. *Curr Biol* 17: 1146-1150.

Leroy, G., 2011. Genetic diversity, inbreeding and breeding practices in dogs: results from pedigree analyses. *The Vet Journal* 189(2): 177-182.

Lickliter, R. and H. Honeycutt. 2010. Chapter 3: Rethinking epigenesis and evolution in the light of developmental science. Pp. 30-47 in M. Blumberg, J. Freeman and S. Robinson (eds.). *Oxford Handbook of Developmental Behavioral Neuroscience.* Oxford University Press.

Lincoln, R., G. Boxshall and P. Clark (eds.). 1998. *A Dictionary of Ecology, Evolution and Systematics,* 2nd ed. Cambridge Univ. Press, Cambridge.

Lindberg, J., et al. 2005. Selection for tameness has changed brain gene expression in silver foxes. *Curr Biol* 15(22): R915-R916.Lindblad-Toh, K., et al. 2005. Genome sequence, comparative analysis and haplotype structure of the domestic dog. *Nature* 438: 803-819.

Linnaeus, C. 1758. *Systema naturae, Vol. 1.*

Liu, C. H. 1932. The dog-ancestor story of the aboriginal tribes of southern China. *J Royal Anthropol Inst GB and Ireland* 62: 361-368.

Liu, Y. P., et al. 2006. Multiple maternal origins of chickens: out of the Asian jungles. *Molecular Phylog Evol* 38(1): 12-19.

Losos, J. B. 1999. Uncertainty in the reconstruction of ancestral states and limitations on the use of phylogenetic comparative methods. *Anim Behav* 58: 1319-1324.

Louys, J. and E. Meijaard. 2010. Palaeoecology of Southeast Asian megafauna-bearing sites from the Pleistocene and a review of environmental changes in the region. *J Biogeogr* 37(8): 1432-1449.

Lyras, G. and A. A. E. Van der Geer. 2003. External brain anatomy in relation to the phylogeny of Caninae (Carnivora: Canidae). *Zool J Linn Soc* 138: 505-522.

Macdonald, D. W. 1979. The flexible social system of the golden jackal, *Canis aureus*. *Behav Ecol Sociobiol* 5(1): 17-38.

Macdonald, D. W. and G. M. Carr. 1995. Variation in dog society: between resource dispersion and social flux. Pp. 199-216 in J. Serpell (ed.), *The Domestic Dog: Its evolution, behaviour and interactions with people*.

Macdonald, D. W. and C. Sillero-Zubiri (eds.). 2004. *The Biology and Conservation of Wild Canids*. Oxford Univ. Press, Oxford.

MacLean, E. L., et al. 2012. How does cognition evolve? Phylogenetic comparative psychology. *Anim Cognition* 15(2): 223-238.

MacNulty, D. R., et al. 2012. Nonlinear effects of group size on the success of wolves hunting elk. *Behav Ecol* 23(1): 75 - 82.

Majumder, A., et al. 2011. Food habits and temporal activity patterns of the Golden Jackal *Canis aureus* and the Jungle Cat *Felis chaos* in Pench Tiger Reserve, Madhya Pradesh. *J Threatened Taxa* 3(11): 2221-2225.

Majumder, S. S., et al. 2013. To be or not to be social: foraging associations of free-ranging dogs in an urban ecosystem. *Acta Ethol* doi: 10.1007/s10211-013-01

Majumder, S. S., A. Chatterjee and A. Bhadra. 2014. A dog's day with humans-time activity budget of free-ranging dogs in India. *Research Commun* 106(6): 874-878.

Malcom, J. R. 1985. Paternal care in canids. *Amer Zool* 25(3): 853-856.

Manwell, C. and C. M. A. Baker. 1983. Origin of the dog: from wolf or wild *Canis familiaris*? *Spec Sci Tech* 6(3): 213-224.

Manwell, C. and C. M. A. Baker. 1984. Domestication of the dog: hunter, food, bed-warmer or emotional object? *Z. Tierzüchtg Züchtgsbiol* 101: 241-256.

Marshall, F. B., et al. 2014. Evaluating the roles of directed breeding and gene flow in animal domestication. *PNAS* 111(17): 6153-6158.

Marshall-Pescini, S. and J. Kaminski. 2014. The social dog: history and evolution. Pp. 3-33 in J. Kaminski and S. Marshall-Pescini (eds.) *The Social Dog*.

Martin, L. D. 1989. Fossil history of terrestrial Carnivora. Pp. 536-568 in J. L Gittleman (ed.), *Carnivore behavior, ecology, and evolution, Vol. 1*.

Martín, P. et al. 2014. Butchered and consumed: Small carnivores from the Holocene levels of El Mirador Cave (Sierra de Atapuerca, Burgos, Spain). *Quat Inter* 2014: 353-153.

Mason, I. L. (ed.). 1984. *The Evolution of Domesticated Animals*. Longman, London, NY.

Matthew, W. D. 1930. The phylogeny of dogs. *J Mammal* 11:117-138.

Martínez-Navarro, B. and L. Rook. 2003. Gradual evolution in the African hunting dog lineage: systematic implications. *Comptes Rendus Palevol* 2: 695-702.

Marvin, G. 2012. *Wolf*. Reaktion Books, London.

Marwick, B. 2009. Biogeography of Middle Pleistocene hominins in mainland Southeast Asia: A review of current evidence. *Quatern Int* 202(1-2): 51-58.

Masuda, R., et al. 1996. Molecular phylogeny of mitochondrial cytochrome b and 12S rRNA sequences in the Felidae: ocelot and domestic cat lineages. *Mol Phylogenet Evol* 6(3): 351-365.

Mayr, E. 1970. *Populations, Species, and Evolution*. Harvard Univ. Press, Cambridge, MA.

Mayr, E. & Ashlock, P. D. 1991. *Principles of Systematic Zoology*. McGraw-Hill, NY.

McDonald Pavelka, M. S. 2007. Review of *Man the Hunted* by D. Hart and R. Sussman. *Int. Primatol* 28(5): 1193-1194.

McLeod, P. J. and J. C. Fentress. 1997. Developmental changes in the sequential behavior of interacting timber wolf pups. *Behav Process* 39: 127-136.

McNay, M. 2002. *A case history of wolf-human encounters in Alaska and Canada*. Wildlife Techincal Bulletin 13, Alaska Dept. of Fish and Game, Juneau, AK.

Meachen, J. A. and J. X. Samuels. 2012. Evolution in coyotes (*Canis latrans*) in response to the megafaunal extinctions. *PNAS* 109(11): 4191-4196.

Mech, L. D. 2000. Leadership in wolf, *Canis lupus*, packs. *Can Field Nat* 114: 259-263.

Mech, L. D. and L. Boitani (eds.). 2003a. *Wolves: Behavior, Ecology, and Conservation*. Univ. of Chicago Press, Chicago, IL.

Mech, L. D. and L. Boitani. 2003b. Wolf social ecology. Pp. 1-34 in L. D. Mech and L. Boitani (eds.), *Wolves: Behavior, Ecology, and Conservation*.

Meek, P. M. 1999. The movement, roaming behaviour and home range of free-roaming domestic dogs, *Canis lupus familiaris*, in coastal New South Wales. *Wildlife Res* 26(6): 847-855.

Meggitt, M. J. 1965. The association of Australian aborigines and dingoes. Pp. 7-26 in *Man, Culture and Animals*, A. Leeds and A. P. Vayada (eds.), AAAS Publication 78, Washington, DC.

Meijaard, E. 2003. Mammals of south-east Asian islands and their Late Pleistocene environments. *J Biogeogr* 30: 1245-1257.

Meloro, C. 2011. Feeding habits of Plio-Pleistocene large carnivores as revealed by the mandibular geometry. *J Vertebr Paleontol* 31(2): 428-446.

Meloro, C. and P. Raia. 2010. Cats and dogs down the tree: the tempo and mode of evolution in the lower carnassial of fossil and living Carnivora. *Evol Biol* 37: 177-186.

Meloro, C. and P. O'Higgins. 2011. Ecological adaptations of mandibular form in fissiped Carnivora. *J Mamm Evol* 18(3): 185-200.

Meloro, C., A. Hudson and L. Rook. 2014. Feeding habits of extant and fossil canids as determined by their skull geometry. *J Zool* doi: 10.1111/jzo.12196

Menzel, R. and R. Menzel. 1948. Observations on the Pariah dog. Pp. 968-990 in B. Vesey-Fitzgerald (ed.), *The Book of the Dog*. Nicholson and Watson, London.

Mersmann, D., et al. 2011. Simple mechanisms can explain social learning in domestic dogs (*Canis familiaris*). *Ethology* 117(8): 675-690.

Miklósi, Á 2007. *Dog Behaviour, Evolution and Cognition*. New York: Oxford University Press.

Miklósi, Á., et al. 2003. A simple reason for a big difference: wolves do not look back at humans, but dogs do. *Curr Biol* 13: 763-766.

Miklósi, Á. and J. Topál. 2005. Is there a simple recipe for how to make friends? *Trends Cogn Sci* 9(10): 463-464.

Mitchell, D., E. Thatcher Beatty and P. K. Cox. 1977. Behavioral differences between two populations of wild rats: implications for domestication research. *Behav Biol* 19: 206-216.

Moehlman, P. D. 1983. Socioecology of silverbacked and golden jackals (*Canis mesomelas* and *Canis aureus*). *Adv Study Mamm Behav* (7): 423-453.

Montagu, M. F. A. 1942. On the origin and domestication of the dog. *Science* 96: 111-112.

Monzón, J., R. Kays, and D. E. Dykhuizen. 2014. Assessment of coyote-wolfdog admixture using ancestry-informative diagnostic SNPs. *Mol Ecol* 23: 182-197.

Moon-Fanelli, A. 2011. The ontogeny of expression of communicative genes in Coyote-Beagle hybrids. *Behav Genetics* 41(6): 858-875.

Moore, J. 2005. Comparing wolf and dog behavior. 7 pp. http://www.northernterritories.com/wolfs_&_hybrids.htm

Morey, D. F. 1992. Size, shape and development in the evolution of the domestic dog. *J Archaeol Sci* 19: 181-204.

Morey, D. F. 1986. Studies on Amerindian dogs: taxonomic analysis of canid crania from the Northern Plains. *J Archaeol Sci* 13: 119-145.

Morey, D. F. and M. D. Wiant. 1992. Early Holocene domestic dog burials from North America. *Curr Anthropol* 33(2): 224-229.

Morey, D. F. 2010. *Dogs: domestication and the development of a social bond*. Cambridge Univ. Press, NY.

Morey, D. F. 2014. In search of Paleolithic dogs: a quest with mixed results. *J Archaeol Sci* 52: 300-307.

Morey, D. F. and K. Aaris-Sørensen. 2005. Paleoeskimo dogs of the Eastern arctic. *Arctic* 55(1): 44-56.

Morrison, D. 1984. A note on Thule culture dogs from Coronation Gulf, N.W.T. *Can J Archaeol* 8(2): 149-157.

Morrison, K. D. and L. L. Junker. 2002. *Forager-traders in South and Southeast Asia*. Cambridge Univ. Press, NY.

Morwood, M. J., et al. 2005. Further evidence for small-bodied hominins from the Late Pleistocene of Flores, Indonesia. *Nature* 437(7061): 1012-1017.

Muñoz-Durán, J. and B. Van Valkenburgh. 2006. The Rancholabrean record of Carnivora: taphonomic effect of body size, habitat breadth, and the preservation potential of caves. *Palaios* 21: 424-430.

Muñoz-Fuentes, V., et al. 2009. Ecological factors drive differentiation in wolves from British Columbia. *J Biogeogr* 36: 1516-1531.

Musiani, M. et al. 2003. Wolf depredation trends and the use of fladry barriers to protect livestock in western North America. *Conserv Biol* 17(6): 1538-1547.

Musiani M. et al. 2007. Differentiation of pilot/taiga and boreal coniferous forest wolves: genetics, coat colour and association with migratory caribou. *Mol Ecol* 16(19): 4149-4170.

Musil, R. 2001. Evidence for the domestication of wolves in Central European Magdalenian sites. Pp. 21-28 in S. J. Crockford (ed.), *Dogs through Time: An Archaeological Perspective*. Nabholz, B., S. Glémin, S. and N. Galtier. 2009. The erratic mitochondrial clock: variations of mutation rate, not population size, affect mtDNA diversity across birds and mammals. *BMC Evol Biol* 9(1): 54 doi: 10.1186/1471-2148-9-54

Naderi, S., et al. 2008. The goat domestication process inferred from large-scale mitochondrial DNA analysis of wild and domestic individuals. *PNAS* 105(46): 17659-17664.

Naish, D. 2006. Controversial origins of the domestic dog. Oct. 6 Tetrapod Zoology. http://

darrennaish.blogspot.com/

Napierala, H. and H-P. Uerpmann. 2010. A 'new' Palaeolithic dog from Central Europe. *Int J Osteoarchaeol* 22(2): 127-137.

Negus, Sir V. 1958. *The Comparative Anatomy and Physiology of the Nose and Paranasal Sinuses*. E & S Livingstone Ltd., Edinburgh.

Newsome A. E., P. C. Catling, and L. K. Corbett. 1983. The feeding ecology of the dingo II. Dietary and numerical relationships with fluctuating prey populations in south-eastern Australia. *Austral J Ecol* 8: 345-366.

Newsome, T. M., et al. 2014a. Human-resource subsidies alter the dietary preferences of a mammalian top-predator. *Oecologia* 175: 139-150.

Newsome, T. M., et al. 2014b. Dietary niche overlap of free-roaming dingoes and domestic dogs: the role of human-provided food. *J Mam* 95(2): 392-403.

Nichols, R. A. 1996. Assessing relatedness and evolutionary divergence: why the genetic evidence alone is insufficient. Pp. 365-379 in T. B. Smith and R. K. Wayne (eds.), *Molecular Genetic Approaches in Conservation*. Oxford Univ. Press, NY.

Nowak, R. M. 1979. *North American Quaternary Canis*. Museum of Natural History, University of Kansas.

Nowak, R. M. 2002. The original status of wolves in eastern North America. *Southeast Nat* 1(2): 95-130.

O'Connor, T. P. 1997. Working at relationships: another look at animal domestication. *Antiquity* 71: 149-156.

O'Connor, S., R. Onon and C. Clarkson. 2011. Pelagic fishing at 42,000 years before the present and the maritime skills of modern humans. *Science* 334: 1117-1121.

Oliva, J. L., et al. 2015. Oxytocin enhances the appropriate use of human social cues by the domestic dog (*Canis familiaris*) in an object choice task. *Anim Cog* 18(3): 767-775.

Ollivier, M., et al. 2013. Evidence of coat color variation sheds new light on ancient canids. *PLOS One* 8(10): e75110. doi:10.1371/journal.pone.0075110

Olsen, S. J. 1985. *Origins of the Domestic Dog: The Fossil Record*. Univ. of Arizona Press, Tucson, Arizona.

Olsen, S. J. and J. W. Olsen. 1977. The Chinese wolf, ancestor of New World dogs. *Nature* 197: 533-535.

Olsen, S. J. and J. W. Olsen. 1982. The position of *Canis lupus variabilis*, from Zhoukoudian, in the ancestral lineage of the domestic dog, *Canis familiaris*. *Vertebrata PalAsiatica* 20(3): 264-267.

Olsen, S. J., J. W. Olsen and Q. Guo-qin. 1980. Domestic dogs from the Neolithic of China. *Explorers Journal* December: 165-167.

O'Neill, A. 2002. *Living with the Dingo*. Envirobook, Annadale, NSW, Australia.

Oppenheimer, E. C. and J. R. Oppenheimer. 1975. Certain behavioral features in the pariah dog (*Canis familiaris*) in West Bengal. *Appl Anim Ethol* 2(1): 81-92.Oskarsson, M. C. R., et al. 2011. Mitochondrial DNA data indicate an introduction through Mainland Southeast Asia for Australian dingoes and Polynesian domestic dogs. *Proc R Soc B* 279: 967-974.

Ostrander, E. A. (ed.) 2012. *Genetics of the Dog*. CABI, Wallingford, UK.

Ostrander, E. A and R. K. Wayne. 2005. The canine genome. *Genome Res* 15: 1706-1716.

Ostrander, E. A., U. Giger and K. Lindblad-Toh (eds.). 2006. *The Dog and Its Genome*. Cold Springs Harbor Laboratory Press, NY.

Ovodov N. D., et al. 2011. A 33,000-year-old incipient dog from the Altai Mountains of Siberia: evidence of the earliest domestication disrupted by the last glacial maximum. *PLoS ONE* 6(7):e22821. doi:10.1371/journal.pone.0022821

Packard, J. M. 2003. Wolf behavior: reproductive, social, and intelligent. Pp. 35-65 in L. D. Mech and L. Boitani (eds.), *Wolves: Behavior, Ecology and Conservation*.

Pal, S. K. 2003. Urine marking by free-ranging dogs (*Canis familiaris*) in relation to sex, season, place and posture. *Appl Anim Behav Sci* 80: 45-59.

Pal, S. K. 2005. Parental care in free-ranging dogs, *Canis familiaris*. *Appl Anim Behav Sci* 90(1): 31-47.

Pal, S. K. 2008. Maturation and development of social behaviour during early ontogeny in free-ranging dog puppies in West Bengal, India. *Appl Anim Behav Sci* 111(1): 95-107.

Pal, S. K. 2010. Play behaviour during early ontogeny in free-ranging dogs (*Canis familiaris*). *Appl Anim Behav Sci* 126: 140-153.

Pal, S. K., B. Ghosh and S. Roy. 1998a. Dispersal behaviour of free-ranging dogs (*Canis familiaris*) in relation to age, sex, season and dispersal distance. *Appl Anim Behav Sci* 61(2): 123-132.

Pal, S. K., B. Ghosh and S. Roy. 1998b. Agonistic behaviour of free-ranging dogs (*Canis familiaris*) in relation to season, sex and age. *Appl Anim Behav Sci* 59: 331-348.

Pal, S. K., B. Ghosh and S. Roy. 1999. Inter-and intra-sexual behaviour of free-ranging dogs (*Canis familiaris*). *Appl Anim Behav Sci* 62(2): 267-278.

Palmqvist, P., B. Martinez-Navarro and A. Arribas. 1996. Prey selection by terrestrial carnivores in a lower Pleistocene paleocommunity. *Paleobiology* 22(4): 514-534.

Palmqvist, P., A. Arribas and B. Martinez-Navarro. 1999. Ecomorphological study of large canids from the lower Pleistocene of southwestern Spain. *Lethaia* 32: 75-88.

Palmqvist, P., et al. 2002. Estimating the body mass of Pleistocene canids: discussion of some methodological problems and a new 'taxon free' approach. *Lethaia* 35: 358-360.

Pang, J.-F., et al. 2009. mtDNA data indicate a single origin for dogs south of Yangtze River, less than 16,300 years ago, from numerous wolves. *Mol. Biol. Evol.* 26 (12): 2849-2864.

Paquet, P. C. 1992. Prey use strategies of sympatric wolves and coyotes in Riding Mountain National Park, Manitoba. *J Mammal* 73 (2): 337-343.

Parker, H. G., et al. 2004. Genetic structure of the purebred domestic dog. *Science* 304: 1160-1164.

Parker, H. G., A. L. Shearin and E. A. Ostrander. 2010. Man's best friend becomes biology's best in show: genome analyses in the domestic dog. *Ann Rev Genetics* 44: 309-336.

Patterson, N., et al. 2006. Genetic evidence for complex speciation of humans and chimpanzees. *Nature* 441: 1103-1108.

Paxton, D. W. 2000. A case for a naturalistic perspective. *Anthrozoös* 13: 5-8.

Pearson, H. 2006. Genetics: what is a gene? *Nature* 441(7092): 398-401.

Pei, W-C. 1934. *On the carnivora from locality 1 of Choukoutien*. Geological Survey of China, Peking

Pei, W.-C. 1936. *On the mammalian remains from locality 3 at Choukoutien*. Cenozoic research Laboratory, Geological Survey of China, Peking.

Perri, A. R., G. M. Smith and M. D. Bosch. 2015. Comment on "How do you kill 86 mammoths? Taphonomic investigations of mammoth megasites" by Pat Shipman. *Quat Int* 368(11): 2e115.

Peterson, R. O. 1977. *Wolf ecology and prey relationships on Isle Royale*. U. S. National Park Service Scientific Monograph Series, no. 11, Washington D. C.

Peterson, R. O., J. D. Woolington and T. N. Bailey. 1984. Wolves of the Kenai peninsula, Alaska. *Wildlife Monographs 88:* 52 pp.

Pilot. M., et al. 2006. Ecological factors influence population genetic structure of European grey wolves. *Mol Ecol.* 14: 4533-53.

Pilot, M., et al. 2015. On the origin of mongrels: evolutionary history of free-breeding dogs in Eurasia. *Proc. R. Soc. B* 2015 282 20152189; DOI: 10.1098/rspb.2015.2189

Pionnier-Capitan, M., et al. 2011. New evidence for Upper Palaeolithic small domestic dogs in South-Western Europe. *J Archaeol Sci* 38: 2123-2140.

Poché, R. M., et al. 1987. Notes on the golden jackal (*Canis aureus*) in Bangladesh. *Mammalia* 51(2): 259-270.

Podberscek, A. L. 2009. Good to pet and eat: The keeping and consuming of dogs and cats in South Korea. *J Soc Issues* 65(3): 615-632.

Polaszek, A. and E. O. Wilson. 2005. Sense and stability in animal names. *Trends in ecology & evolution* 20(8): 421-422.

Polavarapu, N., et al. 2011. Characterization and potential functional significance of human-chimpanzee large INDEL variation. *Mobile DNA* 2: 13 DOI: 10.1186/1759-8753- 2-13

Pongrácz, P., et al. 2001. Social learning in dogs: the effect of a human demonstrator on the performance of dogs in a detour task. *Anim Behav* 62: 1109—1117.

Ptashne, M. 2013. Epigenetics: Core misconcept. *PNAS.* 110 (18): 7101-7103.

Puja, I. K., et al. 2005. The Kintamani dog: genetic profile of an emerging breed from Bali, Indonesia. *J. Heredity* 96(7): 854-859.

Pulquério, M. J. F. and R. A. Nichols. 2007. Dates from the molecular clock: how wrong can we be? *Trends Ecol Evol* 22: 180-184.

Purcell, B. 2010. *Dingo*. CISRO Publ., Collinwood, VIC, Australia.

Radinsky, L. B. 1981a. Evolution of skull shape in carnivores, 1: Representative modern carnivores. *Biological Journal of the Linnean Society* 15(4): 369-388.

Radinsky, L. B. 1981b. Evolution of skull shape in carnivores, 2: Additional modern carnivores. *Biological Journal of the Linnean Society* 16(4): 337-355.

Randi, E. and V. Lucchini. 2002. Detecting rare introgression of domestic dog genes into wild wolf (*Canis lupus*) populations by Bayesian admixture analyses of microsatellite variation. *Conserv Genet* 3(1): 29-43.

Range, F. and Z. Virányi. 2011. Development of Gaze Following Abilities in Wolves (*Canis lupus*). *PLoS ONE* 6(2): e16888. doi:10.1371/journal.pone.0016888

Reed, C. A. 1954. Animal domestication in the Near East. *Science* 130: 1629-1639.

Reponen, S. E., et al. 2014. Genetic and morphometric evidence on a Galápagos Island exposes founder effects and diversification in the first-known (truly) feral western dog population. *Molecular Ecol* 23(2): 269-283.

Rick, T. C., et al. 2009. Origins and antiquity of the island fox (*Urocyon littoralis*) on California's Channel Islands. *Quaternary Research* 71(2): 93-98. Riedel, J., et al. 2008. The early ontogeny of human dog communication. *An Behav* 75 (3): 1003-1014.

Ritchie, E. G., et al. 2013. Chapter 2. Dogs as predators and trophic regulators. Pp. 55-68 in M. Gompper (ed.) *Free-ranging dogs and wildlife conservation.* Oxford University Press.

Roberts, D. 1979. Mechanical structure and function of the craniofacial skeleton of the domestic dog. *Acta Anat* 103: 422-433.

Roberts, T., P. McGreevy and M. Valenzuela. 2010. Human induced rotation and reorganization of the brain of domestic dogs. *PLoS ONE* 5(7): e11946. doi:10.1371/journal.plone.0011946

Rodriguez-Vidal, J., et al. 2014. A rock engraving made by Neanderthals in Gibraltar. *PNAS* doi: 10.1073/pnas.1411529111

Rossel, S. et al. 2008. Domestication of the donkey: timing, processes, and indicators. *PNAS* 205(10): 3715-3720.

Rotter, J. 1999. Der Hahoawu und andere Hunde der tropischen Hackbaukulturen. *Sonderdruck:* 3-24.

Rueness, M., et al. 2011. The cryptic African wolf: *Canis aureus lupaster* is not a golden jackal and is not endemic to Egypt. *PLoS One* 6(1):e16385. doi: 10.1371/journal.pone.0016385

Ruddle, F. H., et al. 1994. Evolution of Hox genes. *Ann Rev Genet* 28(1): 423-442.

Russell, B. 1928. *Sceptical Essays.* W. W. Norton, NY.

Russell, N. 2002. The wild side of animal domestication. *Society & Animals* 10(3): 285-302.

Sablin, M. 2002. Personal message about dog origin date.

Sablin, M. V. and G. A. Khlopachev. 2002. The earliest Ice Age dogs: evidence from Eliseevichi 11. *Curr Anthro* 43(5): 795-799.

Sacks, B. N., et al. 2008. Coyotes demonstrate how habitat specialization by individuals of a generalist species can diversify populations in a heterogeneous ecoregion. *Mol Biol Evol* 25: 1384-1394.

Sacks, B. N., et al. 2013. Y chromosome analysis of dingoes and Southeast Asian village dogs suggests a Neolithic continental expansion from Southeast Asia followed by multiple Austronesian dispersals. *Mol Biol Evol* 30(5): 1103-1118.

Saetre, P., et al. 2004. From wild wolf to domestic dog: gene expression changes in the brain. *Mol Brain Res* 126 (2): 198-206.

Saillard, J. et al. 2000. mtDNA Variation among Greenland Eskimos: the edge of the Beringian expansion. *Am J Hum Genet* 67: 718-726.

Samonte, I. E., et al. 2007. Gene flow between species of Lake Victoria haplochromine fishes. *Mol Biol Evol* 24 (9): 2069-2080.

Sardella, R. and M. R. Palombo. 2007. The Pliocene-Pleistocene boundary: which significance for the so called "wolf event"? Evidences from Western Europe. *Quaternaire* 18(1): 65-71.

Sardella, R., et al. 2013. The wolf from Grotta Romanelli (Apulia, Italy) and its implications in the evolutionary history of *Canis lupus* in the Late Pleistocene of Southern Italy. *Quaternay Inter* http://dx.doi.org/10.1016/j.quaint.2013.11/016

Sathiamurthy, E. and H. K. Voris. 2006. Maps of Holocene sea level transgression and submerged lakes on the Sunda Shelf. *The Natural History Journal of Chulalongkorn University, Supplement* 2: 1-43.

Sato, A., et al. 2011. Spectrum of MHC Class II variability in Darwin's finches and their close relatives. *Mol. Biol. Evol.* 28 (6): 1943-1956.

Saunders, G., et al. 1993. Urban foxes (*Vulpes vulpes*): food acquisition, time and energy budgeting of a generalized predator. *Sym. Zoo. Soc London* 65: 215-234.

Savolainen, P., et al. 2002. Genetic evidence for an East Asian origin of domestic dogs. *Science* 298: 1610-1613.

Savolainen, P., et al. 2004. A detailed picture of the origin of the Australian dingo, obtained from the study of mitochondrial DNA. *PNAS USA* 101: 12387-12390.

Savolainen, P. 2006. mtDNA studies of the origin of the dog. Pp. 119-140 in E. Ostrander, U. Giger and K. Lindblad-Toh (eds.), *The Dog and Its Genome*.

Sawai, H., et al. 2010. The origin and genetic variation of domestic chickens with special reference to jungle fowls *Gallus g. gallus* and *G. varius*. *PLoS ONE* 5(5): e10639. doi:10.1371/journal.pone.0010639

Scally, A. and R. Durbin. 2012. Revising the human mutation rate: implications for understanding human evolution. *Nat Rev Gen* 13(10): 745-753.

Schenekar, T. and S. Weiss. 2011. High rate of calculation errors in mismatch distribution analysis results in numerous false inferences of biological importance. *Heredity* 107: 511-512.

Schenkel, R. 1947. Ausdrucks-Studien an Wölfen: Gefangenschafts-Beobachtungen. *Behaviour* 1(2): 81-129.

Schenkel, R. 1967. Submission: its features and function in the wolf and dog. *Amer Zool* 7 (2): 319-329.

Schirmer, A., C. S. Seow and T. B. Penney. 2013. Humans process dog and human facial affect in similar ways. *PLOS One* 8(9): e74591. doi:10.1371/journal.pone.0074591

Schmitt, E. and S. Wallace. 2012. Shape change and variation in the cranial morphology of wild canids (*Canis lupus, Canis latrans, Canis rufus*) compared to domestic dogs (*Canis familiaris*) using geometric morphometrics. *Int J Osteoarchaeol* 24(1): 42-50.

Schoenebeck, J. J., et al. 2012. Variation of BMP3 contributes to dog breed skull diversity. *PLoS Genet* 8(8): e1002849

Schwartz, M. 1998. *A History of Dogs in the Early Americas*. Yale University Press, New Haven.

Scott, J. P. 1950. The social behavior of dogs and wolves: an illustration of sociobiological systematics. *Annals NY Acad Sci* 51: 1009-1021.

Scott, J. P. 1968. Evolution and domestication of the dog. Pp. 243-275 in T. Dobzhansky, M. K. Hecht and W. C. Steere (eds.), *Evolutionary Biology, Vol. 2*, Plenum Press, NY.

Scott, J. P. and J. L. Fuller. 1965. *Genetics and the Social Behavior of Dogs*. Univ. of Chicago Press.

Scott, J. P., J. H. Shepard and J. Werboff. 1967. Inhibitory training in dogs: effects of age at training basenjis and Shetland sheepdogs. *J Psychol* 66: 237-252.

Scott, M. D. and K. Causey. 1973. Ecology of feral dogs in Alabama. *J Wildlife Manage* 37: 253-265.

Sepkoski, Jr., J. J. 1998. Rates of speciation in the fossil record. *Philos Trans R Soc Lond B Biol Sci* 353(1366): 315-26.

Serpell, J. (ed.). 1995. *The Domestic Dog: Its evolution, behaviour and interactions with people*. Cambridge University Press, Cambridge.

Serpell, J. A., and D. L. Duffy. 2014. Dog breeds and their behavior. Pp. 31-57 in A. Horowitz (ed.), *Domestic Dog Cognition and Behavior*.

Shannon, L. M., et al. 2015. Genetic structure in village dogs reveals a Central Asian domestication origin. *PNAS Early Edition* www.pnas.org/cgi/doi/10.1073/pnas.151621511 2238 Dawn of the Dog

Shaw, J. H. 1975. *Ecology, behavior and systems of the red wolf (Canis rufus)*. Unpublished Ph.D. dissertation, Yale University, New Haven.

Shearin, A. L. and E. A. Ostrander. 2010. Canine Morphology: Hunting for Genes and Tracking Mutations. *PLoS Biol* 8(3): e1000310. doi:10.1371/journal.pbio.1000310

Sheldon, J. W. 1992. *Wild Dogs: the natural history of the Canidae*. Academic Press, San Diego.

Shigehara, N., et al. 1993. First discovery of the ancient dingo-type dog in Polynesia (Pukapuka, Cook Islands). *Int. J. Osteoarchaeology* 3: 315-320.

Shigehara, N. 1994. Morphological changes in Japanese ancient dogs. *Archaeozoologia* V1/2: 79-94.

Shigehara, N., et al. 1998. Morphological study of the ancient dogs from three Neolithic sites in China. *Int. J. of Osteoarchaeology* 8: 11-22.

Shigehara, N. and H. Hongo. 2001. Ancient remains of Jomon dogs from Neolithic sites in Japan. Pp. 61-67 in S. J. Crockford, (ed.), *Dogs through Time: An Archaeological Perspective*.

Shipman, P. 2014. How do you kill 86 mammoths? Taphonomic investigations of mammoth megasites. *Quaternary Int* 359: 38-46.

Shipman, P. 2015a. A reply to Perri et al. (this volume). *Quat International*: 368(11), 6e118.

Shipman, P. 2015b. *The Invaders: how humans and their dogs drove Neanderthals to extinction*. Belknap Press, Harvard Univ. Press, Cambridge, Massachusetts.

Shrotriya, S., S. Lyngdoh and B. Habib. 2012. Wolves in Trans-Himalayas: 165 years of taxonomic confusion. *Cur Sci* 103(8): 885-887.

Sillero-Zubiri, C., M. Hoffmann and D. Macdonald (eds.). 2004. *Canids: foxes, wolves, jackals and dogs*. IUCN, Cambridge, UK.

Skoglund, P., A. Götherström and M. Jakobsson. 2011. Estimation of population divergence times from non-overlapping genomic sequences: examples from dogs and wolves. *Mol Biol Evol* 28 (4): 1505-1517.

Slater, G. J., E. R. Dumont and B. Van Valkenburgh. 2009. Implications of predatory specialization for cranial form and function in canids. *J Zool* 278: 181-188.

Slattery, J. P. and S. J. O'Brien. 1998. Patterns of Y and X chromosome DNA sequence divergence during the Felidae radiation. *Genetics* 148: 1245-1255.

Slik, J. W. F., et al. 2011. Soils on exposed Sunda Shelf shaped biogeographic patterns in the equatorial forests of Southeast Asia. *PNAS* 108 (30): 12343-12347.

Smith, B. P. 2014. Living with wild dogs: personality dimensions in captive dingoes (*Canis dingo*) and implications for ownership. *Anthrozoös* 27(3): 429-433.

Smith, B. and C. A. Litchfield. 2009. A review of the relationship between indigenous Australians, dingoes (*Canis dingo*) and domestic dogs (*Canis familiaris*). *Anthrozoös* 22(2): 111-128.

Smith, A. B. and C. Patterson. 1988. The influence of taxonomic method on the perception of patterns of evolution. *Evol Biol* 23: 127-207.

Smith, B. and C. A. Litchfield. 2010. Dingoes (*Canis dingo*) can use human social cues to locate hidden food. *Anim Cog* 13(2): 367-376.

Smith, B. P. and C. A. Litchfield. 2010. How well do dingoes, *Canis dingo*, perform on the detour task? *Anim Behav* 80(1): 155-162.

Smith, B. P., R. G. Appleby and C. A. Litchfield. 2012. Spontaneous tool-use: An observation of a dingo (*Canis dingo*) using a table to access an out-of-reach food reward. *Behav Process* 89(3): 219-224.

Smukowski, C. S. and M. A. F. Noor. 2011. Recombination rate variation in closely related species. *Heredity* 107: 496-508.Soares, P., et al. 2008. Climate change and postglacial human dispersals in Southeast Asia. *Mol Biol Evol* 25(6): 1209-1218.

Sommer, R. and N. Benecke. 2005. Late-Pleistocene and early Holocene history of the canid fauna of Europe (Canidae). *Mamm Biol* 70(4): 227-241.

Sotnikova, M. V. 2001. Remains of Canidae from the Lower Pleistocene site of Untermassfeld. Pp. 607-632 in R. D. Kahlke (ed.), *Das Pleistozän von bei Meiningen (Thüringen), Vol. 2*. Römisch-Germanisches Zentralmuseum, Bonn.

Sotnikova, M. V. 2002. Personal messages about Eurasian canid paleontology record.

Spalton, A. 2002.Canidae in the Sultanate of Oman. Canid News 5:1 URL:http://www.canids.org/canidnews/5/canids_in_oman.pdf

Speth, J. D. and K. A. Spielmann. 1983. Energy source, protein metabolism, and hunter-gatherer subsistence strategies. *J Anthropol Archaeo* 1: 1-31.

Spotte, S. 2012. *Societies of Wolves and Free-Ranging Dogs*. Cambridge Univ. Press, NY.

Stahl, P. W. 2012. Interactions between humans and endemic canids in Holocene South America. *J Ethnobiology* 32(1): 108-127.

Stahler, D. R. 2002. Interspecific interactions between the common raven (*Corvus corax*) and the gray wolf (*Canis lupus*) in Yellowstone National Park, Wyoming. *Investigations of a predator and scavenger relationship*. Master's thesis, Univ. of Vermont, Burlington.

Stewart, J. B. et al. 2008. Strong purifying selection in transmission of mammalian mitochondrial DNA. *PLoS Biol* 6(1): e10. doi:10.1371/journal.pbio.0060010

Stewart, L., et al. 2015. Citizen science as a new tool in dog cognition research. *PLoS ONE* 10(9): 0135176. doi:10.1371/journal.pone.0135176

Stiner, M. C. 1994. *Honor Among Thieves: A Zooarchaeological Study of Neanderthal Ecology*. Princeton Univ. Press, Princeton, NJ.

Stiner, M. C., N. D. Munro, and T. A. Surovell. 2000. The tortoise and the hare: small-game use, the Broad-Spectrum Revolution, and Paleolithic demography. *Curr Anthro* 41(1): 39- 79.

Stivens, D. 1940. Native dog. *Southerly* 1(2): 24-26.

Storer, T. I. and P. W. Gregory. 1934. Color aberrations in the pocket gopher and their probable genetic explanation. *J Mamm* 15(4): 300-312.

Storz, J. F. and H. E. Hekstra. 2007. The study of adaptation and speciation in the genomic era. *Mammalogy* 88(1): 1-4.

Straus, L. G., et al. (eds.) 1996. *Humans at the End of the Ice Age: The Archaeology of the Pleistocene—Holocene Transition*. Interdisciplinary Contributions to Archaeology. Springer Publ.

Studer T. 1901. Die prähistorischen Hunde in ihrer Beziehung zu den gegenwärtig lebenden Rassen. Zurich: Zurcher und Furrer; pp. 1-154.

Sudarshan, M. K. 2004. *Assessing the burden of rabies in India. WHO sponsored national multi-center rabies survey*. Pp. 44-45. Assn. for the Control of Rabies in India.

Sundqvist, A.-K., H. Ellegren, and C. Vilà. 2008. Wolf or dog? Genetic identification of predators from saliva collected around bite wounds on prey. *Conserv Genet* 9: 1275-1279.

Sutter, N. B., and E. A. Ostrander. 2004. Dog star rising: the canine genetic system. *Nat Rev Genet*

5(12): 900-910. Sutter, N. B., et al. 2007. A single IGF1 allele is a major determinant of small size in dogs. *Science* 316: 112-115.

Suzuki H, et al. 2004. Temporal, spatial, and ecological modes of evolution of Eurasian *Mus* based on mitochondrial and nuclear gene sequences. *Mol Phylogenet Evol* 33(3): 626-46.

Svartberg, K. 2007. Individual differences in behaviour-dog personality. Pp. 182-206 in P. Jensen (ed.) *The Behavioural Biology of Dogs*.

Tang, R., et al. 2014. Candidate genes and functional noncoding variants identified in a canine model of obsessive-compulsive disorder. *Genome Biol* 15(3): R25.

Tanner, J. B., et al. 2008. Of arcs and vaults: the biomechanics of bone-cracking in spotted hyenas (*Crocuta crocuta*). *Biol J Linn Soc* 95: 246-255.

Tchernov, E. and F. F. Valla. 1997. Two new dogs, and other Natufian dogs, from the southern Levant. *J Archaeo Sci* 24: 65-95.

Tedford, R. H., B. E. Taylor and X. Wang. 1995. Phylogeny of the Caninae (Carnivora: Canidae): the living taxa. *Am Mus Novitates* 3146: 1-37.

Tedford, R. H., X. Wang and B. E. Taylor. 2009. Phylogenetic systematics of the North American fossil Caninae (Carnivora: Canidae). *B Am Mus Nat Hist* No. 325. AMNH, Wash. D.C.

Thalmann, O., et al. 2013a. Ancient DNA Analysis Affirms the Canid from Altai as a Primitive Dog. *PLoS ONE* 8(3): e57754. doi: 10.1371/journal.pone.0057754

Thalmann, O., et al. 2013b. Complete mitochondrial genomes of ancient canids suggest a European origin of domestic dogs. *Science* 342: 871-874.

Thompson, J. N. 1998. Rapid evolution as an ecological process. *Trends Ecology Evol* 13(8): 329-332.

Thomson, P. C. 1992a. The behavioural ecology of dingoes in north-western Australia. III. Hunting and feeding behaviour, and diet. *Wildlife Res* 19(5): 531-541.

Thomson, P. C. 1992b. The behavioural ecology of dingoes in north-western Australia. IV. Social and spatial organization, and movements. *Wildlife Res* 19(5): 543-563.

Thomson, P. C. 1992c. The behavioural ecology of dingoes in North-western Australia. II. Activity Patterns, breeding season and pup rearing. *Wildlife Res* 19(5): 519-530.

Thomson, P. C., K. Rose and N. E. Kok. 1992. The behavioural ecology of dingoes in north-western Australia. V. Population dynamics and variation in the social system. *Wildlife Res* 19(5): 565-583.

Titcomb, M. 1969. *Dogs and Man in the Ancient Pacific with Special Attention to Hawaii*. Star-Bulletin Print Co., Honolulu, HI.

Tomasello, M. and B. Hare. 2005. The emotional reactivity hypothesis and cognitive evolution. *Trends Cog Sci* 9(10): 464-465.

Tong, H. W., N. Hu and X. Wang. 2012. New remains of *Canis chihliensis* (Mammalia, Carnivora) from Shanshenmiaozui, a Lower Pleistocene site in Yangyuan, Hebei. *Vertebrata PalAsiatica* 50(4): 335-360.

Topál, J., Á. Miklósi and V. Csányi. 1997. Dog-human relationship affects problem solving behavior in the dog. *Anthrozöos* 10(4): 214-224.

Topál, J., A. Kis and K. Oláh. 2014. Dogs' sensitivity to human ostensive cues: a unique adaptation. Pp. 319-346 in J. Kaminski and S. Marshall-Pecini (eds.), *The Social Dog: Behavior and Cognition*.

Troughton, E. 1957. A new native dog from the Papuan Highlands. *Proc R Zool Soc N S Wales*

1955-1956: 93-94. Troughton, E. 1971. The early history and relationships of the New Guinea Highland dog (*Canis hallstromi*). *Proc Linn Soc N S Wales*. 96: 93-98.

Trut, L.N. et al. 1991. The intracranial allometry and morphological changes in silver foxes (*Vulpes fulvus* dem.) under domestication. *Genetika* 27: 1606-1611.

Trut, L., I. Oskina and A. Kharlamova. 2009. Animal evolution during domestication: the domesticated fox as a model. *BioEssays* 31(3): 349-360.

Tsilioni, I., et al. 2014. Elevated serum neurotensin and CRH levels in children with autistic spectrum disorders and tail-chasing Bull Terriers with a phenotype similar to autism. *Translational Psychiatry* 4(10): e466 doi:10.1038/tp.2014.106

Ucko, P. J. and G. W. Dimbleby (eds.). 1969. *The Domestication and Exploitation of Plants and Animals*. Gerald Duckworth & Co.

Udell, M. A., N. R. Dorey and C. D. Wynne. 2008. Wolves outperform dogs in following human social cues. *Anim Behav* 76(6): 1767-1773.

Udell, M. A. and C. D. Wynne. 2011. Reevaluating canine perspective-taking behavior. *Learning & Behavior* 39(4): 318-323.

Udell, M. A. R., et al. 2014. Exploring breed differences in dogs (*Canis lupus familiaris*): Does exaggeration or inhibition of predatory response predict performance on human-guided tasks? *Anim Behav* 89: 99-105.

University of Illinois at Urbana-Champaign. 2009 (December 30). Transcription factors guide differences in human and chimp brain function. *ScienceDaily*. Retrieved August 6, 2012, from http://www.sciencedaily.com /releases/2009/12/091207151220.htm

Våge, J., et al. 2010. Association of dopamine- and serotonin-related genes with canine aggression. *Genes Brain Behav*. 9(4): 372-378.

Van den Bergh, G. D. 1999. The Late Neogene elephantoid bearing faunas of Indonesia and their palaeozoogeographic implications; a study of the terrestrial faunal succession of Sulawesi, Flores and Java, including evidence for early hominid dispersal east of Wallace's line. *Scripta Geologica* 117: 1-419.

Van der Borg, J. A. M., et al. 2015. Dominance in Domestic Dogs: A Quantitative Analysis of Its Behavioural Measures. *PLoS ONE* 10(8): e0133978. doi:10.1371/journal.pone.0133978

Van Kerkhove, W. 2004. A fresh look at the wolf-pack theory of companion-animal dog social behavior. *J. Appl. Anim Welfare Sci* 7(4): 279-285.

Van Valkenburgh, B. 1989. Carnivore dental adaptations and diet: a study of trophic diversity within guilds. Pp. 410-436 in: J. L. Gittleman (ed.), *Carnivore behavior, ecology, and evolution, Vol. 1*.

Van Valkenburgh, B. 1991. Iterative evolution of hypercarnivory in canids (Mammalia: Carnivora): evolutionary interactions among sympatric predators. *Paleobiology* 17(4): 340-362.

Van Valkenburgh, B. and K. P. Koepfli. 1993. Cranial and dental adaptations to predation in canids. *Symp Zool Soc Lond* 65: 15-37.

Van Valkenburgh, B. and R. K. Wayne. 1994. Shape divergence associated with size convergence in sympatric east African jackals. *Ecology* 75(6): 1567-1581.

Van Valkenburgh, B., X. M. Wang and J. Damuth. 2004. Cope's rule, hypercarnivory, and extinction in North American canids. *Science* 306: 101-104.

Vanak, A. T. and M. E. Gompper. 2009. Dietary niche separation between sympatric free-ranging domestic dogs and Indian foxes in central India. *J Mammal* 90(5): 1058-1065.Vanak, A. T. and M. E. Gompper. 2009. Dogs *Canis familiaris* as carnivores: their role and function in intraguild competition. *Mammal Rev.* 39(4): 265-283.

Varki, A. and T. K. Altheide. 2005. Comparing the human and chimpanzee genomes: Searching for needles in a haystack. *Genome Res.* 15: 1746-1758.

Vergano, D. 2002. Scientists dogged by question of origin. *USA Today* Sept. 30. http://www.usatoday.com/news/health/2002-09-30 -dogsorigins-usat_x.htm

Verginelli, F., et al. 2005. Mitochondrial DNA from Prehistoric Canids Highlights Relationships Between Dogs and South-East European Wolves. *Mol Biol Evol* 22 (12): 2541-2551.

Verheyen, E., et al. 2003. Origin of the superflock of cichlid fishes from Lake Victoria, East Africa. *Science* 300 (5617): 325-329.

Vermeire, S., et al. 2012. Serotonin 2A receptor, serotonin transporter and dopamine transporter alterations in dogs with compulsive behaviour as a promising model for human obsessive-compulsive disorder. *Psychiatry Research: Neuroimaging:* 201(1): 78-87.

Vernes, K., A. Dennis and J. Winter. 2001. Mammalian diet and broad hunting strategy of the Dingo (*Canis familiaris dingo*) in the wet tropical rain Forests of Northeastern Australia. *Biotropica* 33 (2): 339-345.

Vesey-Fitzgerald, B. (ed.). *The Book of the Dog.* Nicholson and Watson, London.

Via, S. 2001. Sympatric speciation in animals: the ugly duckling grows up. *TRENDS in Ecology & Evolution* 16 (7): 381-390.

Via, S. 2012. Divergence hitchhiking and the spread of genomic isolation during ecological speciation-with-gene-flow. *Philos Trans R Soc Lond B Biol Sci.* 367(1587): 451-460.

Vilà, C., et al. 1997. Multiple and ancient origins of the domestic dog. *Science* 276(5319): 1687-1689.

Vilà, C. and R. K. Wayne. 1999. Hybridization between wolves and dogs. *Conserv Biol.* 13(1): 195-198.

Vilà, C., J. E. Maldonado, and R. K. Wayne. 1999. Phylogenetic relationships, evolution, and genetic diversity of the domestic dog. *J Heredity* 90(1): 71-77.

Vilà, C. and J. A. Leonard. 2012. Canid phylogeny and origin of the domestic dog. Pp. 1-10 in E. A. Ostrander and A. Ruvinski (eds.), *The Genetics of the Dog.*

Vinayak, E., H. Harpending and A. R. Rogers. 2005. Genomics refutes an exclusively African origin of humans. *J Human Evol* 49(1): 1-18.

Virányi, Z., et al. 2008. Comprehension of human pointing gestures in young human-reared wolves (*Canis lupus*) and dogs (*Canis familiaris*). *Anim Cognition* 11: 373-387.

Von Holdt, B, M, et al. 2010. Genome-wide SNP and haplotype analyses reveal a rich history underlying dog domestication. *Nature* 464(7290): 898-902.

Von Holdt, B. M. et al. 2011. A genome-wide perspective on the evolutionary history of enigmatic wolf-like canids. *Genome Res.* 21: 1294-1305.

Voth, I. 1988. *Social behavior of New Guinea dingoes (Canis lupus f. familiaris): expressive behavior, social organization and rank relationships.* PhD (Vet.) thesis, Ludwig Maximilian University, Munich.

Walker, D. N. and G. C. Frison. 1982. Amerindian dogs, 3: Studies on Prehistoric wolf/dog hybrids from the northwestern plains. *J Archaeo Sci* 9 (2): 125-172.

Waller, B. M., et al. 2013. Paedomorphic facial expressions give dogs a selective advantage. *PLoS*

ONE 8(12): e82686. doi: 10.1371/journal.pone.0082686

Wang, G-D., et al. 2013. The genomics of selection in dogs and the parallel evolution between dogs and humans. *Nat Comm* 4:1860 doi: 10.1038/ncomms2814Wang, G.-D., et al 2015. Out of southern East Asia: the natural history of domestic dogs across the world. *Cell Research* doi:10.1038/cr.2015.147

Wang, X., et al. 2004. Chapter 2: Ancestry: Evolutionary history, molecular systematics, and evolutionary ecology of the Canidae. Pp. 39-54 in D. W. Macdonald and C. Sillero-Zubiri (eds.), *The Biology and Conservation of Wild Canids*.

Wang, X. and R. H. Tedford. 2007. Chapter 1. Evolutionary history of canids. Pp. 3-20 in P. Jensen (ed.), *The Behavioural Biology of Dogs*.

Wang, X. and R. H. Tedford. 2008. *Dogs: Their Fossil Relatives and Evolutionary History*. Columbia Univ. Press, NY.

Wayne, R.K. 1986. Cranial morphology of domestic and wild canids: the influence of development on morphological change. *Evolution* 40: 243-261.

Wayne, R. K. 1993. Molecular evolution of the dog family. *Trends Genet* 9: 218-224.

Wayne, R. K., N. Lehman and T. K. Fuller. 1995. Conservation genetics of gray wolves. Pp. 399-407 in L. N. Carbyn, S. H. Fritts and D. R. Seip (eds.), *Ecology and Conservation of Wolves in a Changing World*. Canadian Circumpolar Insitute, Alberta, Canada.

Wayne, R. K. and E. A. Ostrander. 1999. Origin, genetic diversity, and genome structure of the domestic dog. *BioEssays* 21(3): 247-257.

Wayne, R. K. and E. A. Ostrander. 2007. Lessons learned from the dog genome. *Trends in Genetics* 23(11): 557-567.

Weckworth, B. V., et al. 2011. Going coastal: shared evolutionary history between coastal British Columbia and Southeast Alaska Wolves (*Canis lupus*). *PLoS ONE* 6, e19582. doi: 10.1371/journal.pone.0019582

Wheeler, Q. D. and R. Meier (eds.) 2000. *Species Concepts and Phylogenetic Theory*. Columbia Univ. Press, NY.

Whitehouse, S. J. O. 1977. The Diet of the Dingo in Western Australia. *Australian Wildlife Research* 4(2): 145-150.

Wilcox, B. and C. Walkowicz. 1995. *The Atlas of Dog Breeds of the World*. TFH Publ., Inc., Neptune, NJ.

Wilde, N. 2000. *Living with Wolfdogs*. Phantom Publ., Santa Clarita, CA.

Wilkins, A. S., R.W. Wrangham and W. Tecumseh Fitch. 2014. The "Domestication Syndrome" in Mammals: A Unified Explanation Based on Neural Crest Cell Behavior and Genetics. *Genetics* 197: 795-808.

Will, U. 1973. Untersuchungen zur taxonomischen Bedeutung des Kleinhirns der Gattung Canis. *J Zool Systematics and Evolutionary Research* 11: 61-73.

Willingham, A. T. and T. R. Gingeras. 2006. TUF love for "junk" DNA. *Cell* 125(7): 1215-1220.

Wilson, D. E. and D. M. Reeder (eds). 2005. *Mammal Species of the World: A Taxonomic and Geographic Reference*. 3rd ed. Volume 1. Johns Hopkins University Press, Baltimore.

Winston, J. E. 1999. *Describing Species: Practical Taxonomic Procedure for Biologists*. Columbia Univ. Press, NY.

Witt, K. E., et al. 2015. DNA analysis of ancient dogs of the Americas: Identifying possible founding haplotypes and reconstructing population histories. *J Human Evol* 79: 105-118.

Wobber, V., et al. 2009. Breed differences in domestic dogs'(*Canis familiaris*) comprehension of human communicative signals. *Interaction Studies* 10(2): 206-224.

Wolpoff, M. H. 1999. *Paleoanthropology*. McGraw-Hill, NY.

Wood-Jones, F. 1931. Cranial characters of the Hawaiian dog. *J Mammal* 12: 39-41.Woodroffe, R., et al. 2007. African Wild Dogs (*Lycaon pictus*) Can Subsist on Small Prey: Implications for Conservation. *J Mamm* 88(1): 181-193.

Worthington, B. E. 2008. An Osteometric Analysis of Southeastern Prehistoric Domestic Dogs. *Electronic Theses, Treatises and Dissertations*. Paper 753. http://diginole.lib.fsu.edu/etd/753

Wroe, S., et al. 2007. Computer simulation of feeding behaviour in the thylacine and dingo as a novel test for convergence and niche overlap. *Proc Roy Soc B-Biol Sci* 274: 2819-2828.

Wroe, S., C. McHenry and J. Thomason J. 2005. Bite club: comparative bite force in big biting mammals and the prediction of predatory behaviour in fossil taxa. *Proc Roy Soc B-Biol Sci* 272: 619-625.

Wroe, S. and N. Milne. 2007. Convergence and remarkably consistent constraint in the evolution of carnivore skull shape. *Evolution* 51-5: 1251-1260.

Wurster, C. M., et al. 2010. Forest contraction in north equatorial Southeast Asia during the Last Glacial Period. *PNAS* 107 (35): 15508-15511.

Yamaguchi, N., et al. 2004. Craniological differentiation between European wildcats (*Felis silvestris silvestris*), African wildcats (*F. s. lybica*) and Asian wildcats (*F. s. ornata*): implications for their evolution and conservation. *Biological J Linn Soc* 83(1): 47-63.

Yli-Renko, M., O. Vesakoski and J. E. Pettay. 2015. Personality-dependent survival in the marine isopod *Idotea balthica*. *Ethology* 121(2): 135-143.

Yoon, C.K. 2009. *Naming Nature: The Clash between Instinct and Science*. W. W. Norton Co., NY.

Zachos, F. E. and J. C. Habel (eds.). 2011. *Biodiversity Hotspots: Distribution and Protection of Conservation Priority Areas*. Springer, NY.

Zeder, M. 2012. The domestication of animals. *J Anthropol Res* 68(2): 161-190.

Zeder, M., et al. (eds.) 2006. *Documenting Domestication: New Genetic and Archaeological Paradigms*. U. Cal. Press, Berkeley.

Zeuner, F. E. 1963. *A History of Domesticated Animals*. Harper & Row, NY.

Zhao, X., et al. 2004. Further evidence for paternal inheritance of mitochondrial DNA in the sheep (*Ovis aries*). *Heredity* 93(4): 399-403.

Zhou, S., et al. 2006. Origin of mitochondrial DNA diversity of domestic yaks. *BMC Evol Biol* 6:73 doi: 10.1186/1471-2148-6-73

Zrzavý, J. and V. Řičánková. 2004. Phylogeny of recent Canidae (Mammalia: Carnivora): relative reliability and utility of morphological and molecular datasets. *Zoologica Scripta* 33: 311-333.

INDEX OF NAMES AND TERMS

Abruzzo studies 123-25, 126, 127
activity budget 124, bi-modal 125
adrenalin 65
Adulyadej, B. (Bhumibol) 107
affiliative 100, 125
African wild dog 35, 131, 138
Africanis 107, 111-114, 184-186
Aggarwal, R. K. (Ramesh) 59
aggression 24, 65, 70, 123, 127; same-sex 117
Albert, F. W. (Frank) 66
Allen, B. L. (Benjamin) 120
alloparental care 121
alpha (dominant pack leader) 98-100
Altai 56-57
amylase gene 62-63
Andreoli, G. 123
Atkins, D. L. (David) 61
Auersperg, A. M. I. (Alice) 96
auroch 53, 140
Australian dingo 22, 52, 62, 96, 119-122, 129, 131, 139, 178-183
Axelsson, E. (Emil) 62, 68

Baker, C. M. A. (Ann) 134
Basenji 29, 67, 92, 170, 171
Beach, F. A. (Frank) 122
behavior pattern, inherited 104-105
Bekoff, M. (Marc) 100, 122, 125
Bering strait 61, 78, 205
biological imperative 22, 23, 110
Boitani, L. (Luigi) 101, 102, 106, 123, 124
Bonanni, R. (Roberto) 126, 127
Border Collie 103
Birney, E. (Ewan) 85
Bradley, D. G. (Daniel) 79, 85,
Bradshaw, J. W. S. (John) 129
brain, dog 58-62
breeding seasons 66-67
Brisbin, Jr., I. L. (Lehr) 117
bulla:bullae 41, 44-45

bush dog 59, 141
Butler, J. R. A. (James) 111, 112

Cafazzo, S. (Simona) 126, 127
camel 140
Canaan dog 171, 191-194
Canine Behavioral Assessment and Research Questionnaire (C-BARQ) 97
Canis chihliensis 78, 136
Canis lupus lupaster 161
carrion 102, 113, 144
cauda recurva 49
Central African dog 187-190
child lifting 17
clade 75-78, 80, 151
Clutton-Brock, J. (Juliet) 146
cognition 92-93, 94
Colombeau, G. (Guillaume) 57
commensal 100, 107, 142, 147, 151, 169
confidence interval 81, 89
Coppinger, R and L. (Raymond; Lorna) 29
coprophagy 67-68
Coquerelle, M. (Michael) 57
Corbett, L. (Laurie) 131, 141
Cox, M. P. (Murray) 89
coyote 35, 39, 61, 65, 74, 77-78, 88, 91, 102, 104, 115, 125, 139
Crisler, L. (Lois) 20
Ciucci, P. (Paolo) 106

Darwin's finches 87
den, natal 121, 124, 128
Denisovian 150
detour task 96
dhole 29, 35, 62, 137, 141, 149
Dillon, L. S. (Lawrence) 60-61
dire wolf 78
disuse: hypothesis 45, 58, 152, 158
DNA analysis 36, 71-85; mutation rate 74, 76-77, 79, 81-83, 88-89, 160; dating by 77-81

Dog origin models: Pet Hypothesis 14; Self-domestication 15-16; Natural Species 133-153
domestication effects: breeding season 29, 66-67; morphology 49-52; behavior 103-104
Drake, A. G. (Abby) 41, 57
Duffy, D. L. (David) 97
Durbin, R. (Richard) 88

Eliseevichi 55, 58
Encyclopedia of DNA Elements project (ENCODE) 85
Epstein, H. (Hellmut) 8, 135
ethology 104
Evo Devo 82

facultative pack predator 115, 119, 142
feces, nutrients 113
Fiennes, R. and A. (Richard; Alice) 17, 29-30
Formosan Mountain Dog 195-199
founder number, population 71, 80
Fox, M. (Michael) 8
fox, Russian tame 64-65, 67, 70
Francisci, F. 123
Frank, H. and M. G. (Harry: Martha) 93
Freedman, A. H. (Adam) 80
frontal sinus 42-43
Fuller, J. L. (John) 92, 94

Gajdon, G. K. (Gyula) 96
Galton, F. (Francis) 15
Gásci, M. (Márta) 93
Geist, V. (Valerius) 18
gene definition 86; trees 74-75, 86; expression differences 62, 83, 86, 100
genetic introgression 81, chimerism 83
Germonpré, M. (Mietje) 54, 57
Ghosh, B. 122
Goodman, P. (Pat) 68
Goyet 58
Graves, W. N. (Will) 17-18

Hall, S. S. (Stephen) 85
Hecht, J. (Julie) 91, 97

Hemmer, V-H. 59-60
Herre, W. (Wolf) 58, 60
Hetts, S. (Suzanne) 130
homeobox genes 83-84
Homo erectus 28, 150
Homo sapiens 132, 150
Horowitz, A. (Alexandra) 91, 97
hybrid wolf/dog 39, 85, 88; coyote/dog 39, 104-105; dingo/domestic dog 120, 139, 183
hybridization 29, 30, 36, 39, 75, 87,
hypothesis, scientific definition 153-154

Ice Age 56, 136, 140, 149
ICZN/International Commission on Zoological Nomenclature 37
Indog 108-111, 200-204
Inuit dog 205-210
jackal, golden 60, 74, 89, 125, 138, 144
jackal, black-backed 60
junk DNA 82

Kaminski, J. (Juliane) 94
kangaroo 22, 32, 119
Kershenbaum, A. (Arik) 161, 165
Kis, A. (Anna) 96
Kruska, D. (Dieter) 59, 60
Künzl, C. (Christine) 118
Kurtén, B. (Björn) 135

Lake Victoria cichlid fish 87, 151
Larson, G. (Greger) 79, 80, 85
Lawrence, B. (Barbara) 134
leadership in dog groups 125
LeBoeuf, B. J. (Burney) 122
Leonard, J. A. (Jennifer) 57
Linnaeus, C. (Carl) 34-35
Litchfield, C. A. (Carla) 96
Lucchini, V. (Vittorio) 88
Lyras, G. (George) 61

Macdonald, D. W. (David) 123, 126
mandible, dog and wolf 45-46
Manwell, C. (Clyde) 8, 134
Marshall-Pescini, S. (Sarah) 94

mate preference 110, 120, 122, 131
McNay, M. (Mark) 18
Mech, L. D. (David) 101, 102
megafauna 56, 57, 143
midden 26, 69
Miklósi, A. (Adam) 94
Molecular Clock hypothesis 79
Moon-Fanelli, A. (Alice) 105
Morey, D. (Darcy) 37, 41, 54, 58
mouse, house 142
mutation rate 77-79, 80, 81, 88, 160

Naderi, S. (Saeid) 75
Neanderthal 28; *Homo neanderthalensis* 150
neophilic 118
neotenous 40
neurotransmitters 65-66, 103
New Guinea dingo 24, 32, 52, 138, 173-177; behavior 116-118; howls 161, 163; rope pull test 94; tail 50
Newsome, T. (Thomas) 120
niche 11, 15, 19, 25, 26, 39, 44, 46, 59, 87, 101, 107, 137
norepinephrine 65
Nott, H. M. R. (Helen) 129
nuclear DNA 74, 81-82, 100

Occam's razor 154
Oláh, K. (Katalin) 96
omnivore 45, 132, 142
orbital angle 41
Ortolani, A. (Alessia) 106
Ostrander, E. (Elaine) 71

paedomorphic 41, 46, 82
paedomorphism 41, 158
Pang, J.-F. (Jun-Feng) 80
Pal, S. (Sunil) 108-110
Paleolithic era 26-27, 30, 53; skulls 25, 56-58; dogs 58, 135
paternal care 110-111
Pleistocene 57, 135, 143, 147, 149
Polavarapu, N. (Nalini) 86
pigs 28, 59, 66, 67

Předmostí 55
Przewalski's horse 140
proto-dogs 15, 21, 22, 32, 43, 44, 80

radio-collar 112, 113, 123
Ramadevi, J. (Jetty) 59
Randi, E. (Ettore) 88
ravens 102, 145
red jungle fowl 140
regurgitation 110-111, 121
Rome studies 123, 125-126
rope pull test 94-95
Rueness, E. K. (Eli) 89
Russian tame foxes 64, 65, 67, 70, 92

Saetre, P. (Peter) 65
Savolainen, P. (Peter) 77, 79
Scally, A. (Aylwyn) 88
Schenkel, R. (Rudolf) 98
scientific name 34-35; attribution 35
Schmidt, L. (Lori) 68
Scott, J. P. (John) 14, 23, 92, 94
Schwartz, M. (Marion) 30
sensitive period in socialization 92
serotonin 65
Serpell, J. A. (James) 97
Singh, L. (Lalji) 59
Skoglund, P. (Pontus) 81
Smith, B. P. (Bradley) 96
social behavior, natural dogs 106-121; wolf 98-101
source-sink process 147
species concepts 33; diagnostic traits 34, 38, 41, 49; type specimen 35; type species 35; sister species 133, 152
speciation, sympatric 34, 151; allopatric 34, 38; ecological 138
Studer, T. 8, 134
subspecies 35, 38-39, 56, 72, 74, 76, 85
Sundaland 149, 162
survival rate, puppy 113, 122, 128
tarpan 140
taxonomy 34, 60
Taylor, B. E (Beryl) 136

Tedford, R. H. (Richard) 136
teeth, dog and wolf 14, 15, 25, 37, 39, 44, 56, 59-60
TMRCA 89
Terrill, C. (Ceiridwen) 70
Thalman, O. (Olaf) 59
theory, definition 153-154
Thomson, P. C. 119
Topál, J. (József) 96

Van der Geer, A. A. E. (Alexandra) 61
Vilá, C. (Carles) 76
von Bayern, A. M. P. (Auguste) 96
von Holdt, B. M. (Bridget) 85, 88

Walkowicz, C. (Chris) 116
Wang, G-D. 80, 163
Wang, X. (Xioming) 136
Wayne, R. K. (Robert) 40, 71, 75, 89
Wilcox, B. (Bonnie) 116
Will, U. (Ursula) 61
wolf pack formation 98-99; size 101; seasonal behavior 101
Worthington, B. E. 54

Yates, B. (Bonnie) 39, 44
Zeuner, F. E. (Frederick) 8, 134
Xoloitzcuintli 171

ABOUT THE AUTHOR

Jan with 5 Rhodesian Ridgebacks

Janice Koler-Matznick has degrees in biology and environmental science. She has made a special study of animal behavior concentrating on canids. She is certified as an Applied Animal Behaviorist by the Professional Certification Board of the Animal Behavior Society, and is a member of the Canid Specialist Group of the International Union for the Conservation of Nature. She was a founder of the Primitive and Aboriginal Dog Society and the New Guinea Singing Dog Conservation Society. For 45 years her hobby has been keeping, breeding, and showing Rhodesian Ridgebacks under the kennel name Kandu. Janice lives in Oregon with her husband, Darwin Matznick, a varying array of Ridgebacks, and, for 37 years, an Amazon parrot named Bird.

Additional information and discussion about the origin of the dog are available at: https://www.facebook.com/groups/originofthedog/

CPSIA information can be obtained
at www.ICGtesting.com
Printed in the USA
FSOW03n1252150616
21586FS